QUALITATIVE RESEARCH DESIGN

Second Edition

Applied Social Research Methods Series
Volume 41

APPLIED SOCIAL RESEARCH METHODS SERIES

Series Editors

LEONARD BICKMAN, Peabody College, Vanderbilt University, Nashville
DEBRA J. ROG, Vanderbilt University, Washington, DC

1. **SURVEY RESEARCH METHODS (Third Edition)** by FLOYD J. FOWLER, Jr.
2. **SYNTHESIZING RESEARCH (Third Edition)** by HARRIS COOPER
3. **METHODS FOR POLICY RESEARCH** by ANN MAJCHRZAK
4. **SECONDARY RESEARCH (Second Edition)** by DAVID W. STEWART and MICHAEL A. KAMINS
5. **CASE STUDY RESEARCH (Second Edition)** by ROBERT K. YIN
6. **META-ANALYTIC PROCEDURES FOR SOCIAL RESEARCH (Revised Edition)** by ROBERT ROSENTHAL
7. **TELEPHONE SURVEY METHODS (Second Edition)** by PAUL J. LAVRAKAS
8. **DIAGNOSING ORGANIZATIONS (Second Edition)** by MICHAEL I. HARRISON
9. **GROUP TECHNIQUES FOR IDEA BUILDING (Second Edition)** by CARL M. MOORE
10. **NEED ANALYSIS** by JACK McKILLIP
11. **LINKING AUDITING AND META EVALUATION** by THOMAS A. SCHWANDT and EDWARD S. HALPERN
12. **ETHICS AND VALUES IN APPLIED SOCIAL RESEARCH** by ALLAN J. KIMMEL
13. **ON TIME AND METHOD** by JANICE R. KELLY and JOSEPH E. McGRATH
14. **RESEARCH IN HEALTH CARE SETTINGS** by KATHLEEN E. GRADY and BARBARA STRUDLER WALLSTON
15. **PARTICIPANT OBSERVATION** by DANNY L. JORGENSEN
16. **INTERPRETIVE INTERACTIONISM (Second Edition)** by NORMAN K. DENZIN
17. **ETHNOGRAPHY (Second Edition)** by DAVID M. FETTERMAN
18. **STANDARDIZED SURVEY INTERVIEWING** by FLOYD J. FOWLER, Jr. and THOMAS W. MANGIONE
19. **PRODUCTIVITY MEASUREMENT** by ROBERT O. BRINKERHOFF and DENNIS E. DRESSLER
20. **FOCUS GROUPS** by DAVID W. STEWART and PREM N. SHAMDASANI
21. **PRACTICAL SAMPLING** by GART T. HENRY
22. **DECISION RESEARCH** by JOHN S. CARROLL and ERIC J. JOHNSON
23. **RESEARCH WITH HISPANIC POPULATIONS** by GERARDO MARIN and BARBARA VANOSS MARIN
24. **INTERNAL EVALUATION** by ARNOLD J. LOVE
25. **COMPUTER SIMULATION APPLICATIONS** by MARCIA LYNN WHICKER and LEE SIGELMAN
26. **SCALE DEVELOPMENT** by ROBERT F. DeVELLIS
27. **STUDYING FAMILIES** by ANNE P. COPELAND and KATHLEEN M. WHITE
28. **EVENT HISTORY ANALYSIS** by KAZUO YAMAGUCHI
29. **RESEARCH IN EDUCATIONAL SETTINGS** by GEOFFREY MARUYAMA and STANLEY DENO
30. **RESEARCHING PERSONS WITH MENTAL ILLNESS** by ROSALIND J. DWORKIN
31. **PLANNING ETHICALLY RESPONSIBLE RESEARCH** by JOAN E. SIEBER
32. **APPLIED RESEARCH DESIGN** by TERRY E. HEDRICK, LEONARD BICKMAN, and DEBRA J. ROG
33. **DOING URBAN RESEARCH** by GREGORY D. ANDRANOVICH and GERRY RIPOSA
34. **APPLICATIONS OF CASE STUDY RESEARCH** by ROBERT K. YIN
35. **INTRODUCTION TO FACET THEORY** by SAMUEL SHYE and DOV ELIZUR with MICHAEL HOFFMAN
36. **GRAPHING DATA** by GARY T. HENRY
37. **RESEARCH METHODS IN SPECIAL EDUCATION** by DONNA M. MERTENS and JOHN A. McLAUGHLIN
38. **IMPROVING SURVEY QUESTIONS** by FLOYD J. FOWLER, Jr.
39. **DATA COLLECTION AND MANAGEMENT** by MAGDA STOUTHAMER-LOEBER and WELMOET BOK VAN KAMMEN
40. **MAIL SURVEYS** by THOMAS W. MANGIONE
41. **QUALITATIVE RESEARCH DESIGN (Second Edition)** by JOSEPH A. MAXWELL
42. **ANALYZING COSTS, PROCEDURES, PROCESSES, AND OUTCOMES IN HUMAN SERVICES** by BRIAN T. YATES
43. **DOING LEGAL RESEARCH** by ROBERT A. MORRIS, BRUCE D. SALES, and DANIEL W. SHUMAN
44. **RANDOMIZED EXPERIMENTS FOR PLANNING AND EVALUATION** by ROBERT F. BORUCH
45. **MEASURING COMMUNITY INDICATORS** by PAUL J. GRUENEWALD, ANDREW J. TRENO, GAIL TAFF, and MICHAEL KLITZNER
46. **MIXED METHODOLOGY** by ABBAS TASHAKKORI and CHARLES TEDDLIE
47. **NARRATIVE RESEARCH** by AMIA LIEBLICH, RIVKA TUVAL-MASHIACH, and TAMAR ZILBER
48. **COMMUNICATING SOCIAL SCIENCE RESEARCH TO POLICY-MAKERS** by ROGER VAUGHAN and TERRY F. BUSS
49. **PRACTICAL META-ANALYSIS** by MARK W. LIPSEY and DAVID B. WILSON

QUALITATIVE RESEARCH DESIGN

An Interactive Approach

Second Edition

Joseph A. Maxwell

Applied Social Research Methods Series
Volume 41

For information:

Sage Publications, Inc.
2455 Teller Road
Thousand Oaks, California 91320
E-mail: order@sagepub.com

Sage Publications Ltd.
1 Oliver's Yard
55 City Road
London EC1Y 1SP
United Kingdom

Sage Publications India Pvt. Ltd.
B-42, Panchsheel Enclave
Post Box 4109
New Delhi 110 017 India

Printed in the United States of America

Library of Congress Cataloging-in-Publication Data

Maxwell, Joseph Alex, 1941—Qualitative research design: An interactive approach / Joseph A. Maxwell.—2nd ed.
p. cm.—(Applied social research methods series; no. 41)
Includes bibliographical references and index.
ISBN 0-7619-2607-0 (cloth) — ISBN 0-7619-2608-9 (pbk.)
1. Research—Methodology. I. Title. II. Series.
Q180.55.M4M39 2005
001.4′2—dc22 2004013006

04 05 06 07 10 9 8 7 6 5 4 3 2 1

Acquisitions Editor: Lisa Cuevas Shaw
Editorial Assistant: Margo Beth Crouppen
Production Editor: Melanie Birdsall
Copy Editor: Mattson Publishing Services, LLC
Typesetter: C&M Digitals (P) Ltd.
Proofreader: Cheryl Rivard
Indexer: Sylvia Coates
Cover Designer: Janet Foulger

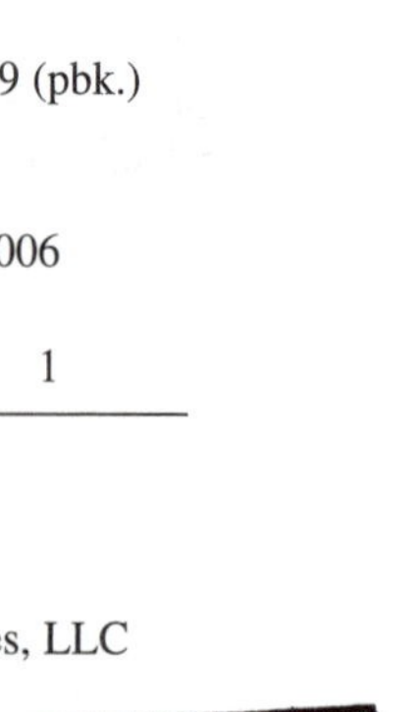

Contents

Preface

I'm very gratified by the positive reception that the first edition of this book has had. The book has received (mostly) very favorable reviews, I've gotten unsolicited "fan mail" from readers who found it valuable, and other authors have adapted and extended some of the ideas I presented.

In this edition, I've made several significant changes. I've added, or substantially expanded my discussion of, a number of topics, including paradigm issues, personal goals, the research problem, research relationships, site and participant selection, data analysis, and validity. I've included additional extended examples in Chapters 2, 3, and 7, and several new exercises, and have revised most of the other exercises, based on my experience with using these in courses and workshops. Finally, I've changed some of the terminology that readers found unclear or that didn't communicate what I intended, and I've compulsively edited the entire text to try to make the language clearer and the organization easier to follow.

One new feature of this edition is that I'm more explicit about the realist perspective that informs this book. Although some of my previous work on research methods was explicitly grounded in a realist philosophical position (e.g., Maxwell, 1992), when I originally wrote this book I wasn't aware of how my approach to qualitative research in general was shaped by realist ideas. Since then, I have done a fair amount of work on the implications of philosophic realism for research methods and social theory (Maxwell, 1999, 2002, 2004a, 2004c), and have come to realize that the model of qualitative design I present here is significantly informed by realist assumptions. This awareness has been aided by a number of recent books that take a realist approach to research methods (Pawson & Tilley, 1997; Robson, 2002; Sayer, 1992). This change is not so much a "coming out of the closet" as a realist (e.g., Robson, 2002, p. ix) as it is discovering (like the character in Molière's play *Le Bourgeois Gentilhomme* who was delighted to discover that all his life he had been speaking prose) that for a long time I had been "thinking realism" without being fully aware of this. The elaboration of this perspective and its implications for qualitative research are the topic for another book; in this one, I simply point out some important consequences of a realist perspective for qualitative research design.

I want to acknowledge and thank all of the doctoral students at George Mason University who have used this book and given me valuable feedback on it, and in particular, the students and former students who have allowed me to use their work as examples. I also want to thank my previous editors at

Sage, C. Deborah Laughton and Heidi Middlesworth, and my current editors, Lisa Cuevas Shaw, Margo Beth Crouppen, and Melanie Birdsall, for their valuable assistance in creating this revision, and for their forbearance when I kept missing deadlines.

Preface to the First Edition

> *Contrary to what you may have heard, qualitative research designs do exist.*
>
> —Miles & Huberman, 1994, p. 16

A review of a book on wines that I once read stated that "A guidebook is best when those guiding you are opinionated," and I have attempted to meet that standard. This book is not a lowest-common-denominator compendium of conventional wisdom. I have definite views on qualitative research design, and a major part of my motivation for writing this book is that I believe that much of what has been written about research design is inconsistent with the way qualitative researchers actually go about designing their research.

I have therefore tried to present an approach to qualitative research design that both captures what qualitative researchers really do, and provides support and guidance for those embarking for the first time on designing a qualitative study. This book is intended to provide advice on every part of the design process: figuring out what your study should accomplish, constructing a theoretical framework, developing research questions, deciding on your strategies and methods for data collection and analysis, and planning how to deal with potential validity threats to your conclusions.

Many students (and some experienced researchers) believe that their research *proposal,* rather than the design of their study, is the major hurdle they face in preparing for their dissertations or in seeking funding. They see qualitative research design as largely common sense, something that doesn't require systematic planning or new ways of thinking. Writing a successful proposal, on the other hand, is viewed as a mysterious and difficult task, requiring them to find the right formula or language that will get their study accepted or funded.

As a result of teaching a course on qualitative research design and proposal writing for the past seven years, I have come to believe the opposite: that *design* is by far the more difficult of the two tasks. Most students in my course discover that they need to substantially rethink and develop their research designs in order to make these consistent and workable, and that this process can force them to re-examine some of their basic assumptions about their topic and study. On the other hand, once they have worked out their research designs more clearly, writing a proposal is not nearly as intimidating or difficult a task as they had feared.

There is a folktale about a boy who was driving a cartload of apples to market. As he had never been to this market before, he asked an old man standing by the road how long it would take to get there. The old man looked at the cart, then looked at the road, and finally said, "About three hours, but if you hurry, it will take you all day." The boy, thinking to himself "What does he know!" whipped up the horse and set off at a fast pace. Soon the cart hit a bump in the road, and many of the apples were knocked out. It took a while for the boy to pick them all up, and because he felt he was late, he now drove even faster. Soon the cart hit another bump, even more apples fell out, and proceeding in this way, it took the boy all day to get to market.

The moral of this story is that if you rush into writing your proposal before you have your design clearly conceptualized, it may substantially slow your progress.[1] Peters (1992, pp. 196–197) provides a cautionary tale of a student who attempted this and wasted a year's worth of work; he comments that the student "should have realized that the reason he couldn't create a clear proposal was that there were basic flaws with the conception of his thesis."

Successfully developing both your research design and your proposal requires that you have a clear understanding of the difference between the two. Your *design* is the logic and coherence of your research study—the components of your research and the ways in which these relate to one another. Your proposal, on the other hand, is a *document* that communicates and justifies this design to a particular audience. Creating these two things are different tasks, and treating them as if they were identical, as some books on research design do, can make both design and proposal writing more difficult than necessary.

For these reasons, this book focuses primarily on research design, rather than on proposal writing. However, there are important connections between design and proposal structure. I highlight these connections throughout the book, and in Chapter 7, I discuss how to make the transition from *designing* a study to *proposing* it. The model of research design that this book presents is partly drawn from the structure of qualitative research proposals, and is thus directly useful in proposal writing.

I've written this book to be useful both to people who are just beginning to plan a qualitative study, and to those who are already involved in a qualitative research project. If you're thinking about or preparing for a qualitative study, you can use the book to develop a research design and proposal. If you're already engaged in a qualitative study, you can use it to reconceptualize what you're doing, focus the study, identify potential problems and solutions to these, and develop more useful and relevant theory, i.e., to modify your design or to make this design more explicit.

I believe that research design, like most things, is best learned by doing it, and I've tried to incorporate a "hands-on" approach to design in this book. Thus, in order for the book to be most useful to you, you should have a qualitative research project in mind that you're either planning to do or are now engaged in. You don't need to have the details of your study worked out—that's what this book should help you to do—but you do need to have a definite topic or subject and some idea of the relevant research and theory on this topic.

The model of design that I present here derives primarily from my experience in teaching qualitative research methods for ten years at the Harvard Graduate School of Education; from the six years that I was a member of the school's Committee on Degrees, which reviews all qualifying paper and dissertation proposals submitted by doctoral students; and from what I have learned in supervising qualitative studies and helping students with their proposals. I've also drawn insight and examples from my own research and proposal writing. My original training was in social and cultural anthropology, at the University of Chicago, but my work has covered a wide range of approaches and topics, from traditional ethnography to qualitative program evaluation and applied research on medical education.

There are far too many people who have contributed to the writing of this book to thank individually, but I particularly want to acknowledge:

Carol Weiss, who first suggested that I write this book, and has been a steady supporter during the writing process.

The students who, over the past seven years, have worked with me as Teaching Fellows in a course on qualitative research design: Maria Broderick, Ana Maria Garcia Blanco, Isabel Londoño, Barbara Miller, Carla Rensenbrink, Anna Romer, Clarissa Sawyer, Janie Simmons, Rachel Sing, Marydee Spillett, and Connie Titone. Most of what is in this book has emerged from or been substantially reshaped by our weekly discussions of research design and students' difficulties in learning this. In addition, many of them read early portions or drafts of the book and provided critical feedback on what worked and what didn't.

Martha Regan-Smith, for permission to use her dissertation proposal as an example, and Suman Bhattacharjea, Maria Broderick, Brendan Croskery, Beatrice Guilbault, Gail Lenehan, Isabel Londoño, Jane Margolis, and Bobby Starnes for permission to use examples from their research in this book.

Loren Faibisch, Beatrice Guilbault, Michael Huberman, Susan Moore Johnson, Matthew Miles, Richard Murnane, Carol Pelletier, Annie Rogers, Ellen Snee, Meg Turner, Robert Weiss, and anyone whom I've forgotten to name, who gave me useful comments on earlier drafts.

All of the students in my course on qualitative research design, who've been the guinea pigs for earlier versions of the presentations and exercises included here, and who have given me a great deal of valuable feedback on these materials.

Meg Turner, for suggesting the subtitle "An Interactive Approach."

C. Deborah Laughton at Sage Publications, and the editors for this series, Len Bickman and Debra Rog, for valuable feedback, encouragement, and prodding to stop revising the book and get it out the door.

Charity Boudouris, for uncomplainingly copying endless drafts of this book to distribute to students and colleagues.

Helen Silver, for preparing the index.

NOTE

1. I am not repeating the fallacy that you should have your argument worked out in your head before you put it on paper; see Becker (1986) for an eloquent refutation of this approach. Working out your design will involve a substantial amount of writing, since, as Becker points out, writing is thinking. However, much of it is a different *sort* of writing from what you will actually put in the proposal; trying to begin by writing something that will persuade an audience of critical reviewers can seriously interfere with the sorts of thinking you need to do to design your study.

1

A Model for Qualitative Research Design

In 1625, Gustav II, the king of Sweden, commissioned the construction of four warships to further his imperialistic goals. The most ambitious of these ships, named the *Vasa,* was one of the largest warships of its time, with 64 cannons arrayed in two gundecks. On August 10, 1628, the *Vasa,* resplendent in its brightly painted and gilded woodwork, was launched in Stockholm Harbor with cheering crowds and considerable ceremony. The cheering was short-lived, however; caught by a gust of wind while still in the harbor, the ship suddenly heeled over, foundered, and sank.

An investigation was immediately ordered, and it became apparent that the ballast compartment had not been made large enough to balance the two gundecks that the king had specified. With only 121 tons of stone ballast, the ship lacked stability. However, if the builders had simply added more ballast, the lower gundeck would have been brought dangerously close to the water; the ship lacked the buoyancy to accommodate that much weight.

In more general terms, the *design* of the *Vasa*—the ways in which the different components of the ship were planned and constructed in relation to one another—was fatally flawed. The ship was carefully built, meeting all of the existing standards for solid workmanship, but key characteristics of its different parts—in particular, the weight of the gundecks and ballast and the size of the hold—were not compatible, and the interaction of these characteristics caused the ship to capsize. Shipbuilders of that day did not have a general theory of ship design; they worked primarily from traditional models and by trial and error, and had no way to calculate stability. Apparently, the *Vasa* was originally planned as a smaller ship, and was then scaled up, at the king's insistence, to add the second gundeck, leaving too little room in the hold (Kvarning, 1993).

This story of the *Vasa* illustrates the general concept of design that I am using here: "an underlying scheme that governs functioning, developing, or unfolding" and "the arrangement of elements or details in a product or work of art" (*Webster's Ninth New Collegiate Dictionary*, 1984). This is the ordinary, everyday meaning of the term, as illustrated by the following quote from a clothing catalog:

> It starts with design. . . . We carefully consider every detail, including the cut of the clothing, what style of stitching works best with the fabric, and what kind of closures make the most sense—in short, everything that contributes to your comfort. (L.L.Bean, 1998)

A good design, one in which the components work harmoniously together, promotes efficient and successful functioning; a flawed design leads to poor operation or failure.

Surprisingly, most works dealing with *research* design use a different conception of design: "a plan or protocol for carrying out or accomplishing something (esp. a scientific experiment)" (*Webster's Ninth New Collegiate Dictionary*, 1984). They present design as a series of stages or tasks in planning or conducting a study. While some versions of this view of design are circular and recursive (e.g., Marshall & Rossman, 1999, pp. 26–27), all are essentially linear in the sense of being a one-directional *sequence* of steps from problem formulation to conclusions or theory, though this sequence may be repeated. Such models usually resemble a flowchart, with a clear starting point and goal and a specified order for performing the intermediate tasks.[1]

Such sequential models are not a good fit for qualitative research, in which any component of the design may need to be reconsidered or modified during the study in response to new developments or to changes in some other component. In a qualitative study, "research design should be a reflexive process operating through every stage of a project" (Hammersley & Atkinson, 1995, p. 24). The activities of collecting and analyzing data, developing and modifying theory, elaborating or refocusing the research questions, and identifying and addressing validity threats are usually all going on more or less simultaneously, each influencing all of the others. This process isn't adequately represented by a linear model, even one that allows multiple cycles, because in qualitative research there isn't an unvarying order in which the different tasks or components must be arranged.

Traditional, linear approaches to design provide a model *for* conducting the research—a prescriptive guide that arranges the tasks involved in planning or conducting a study in what is seen as an optimal order. In contrast, the model in this book is a model *of* as well as *for* research. It is intended to help you understand the *actual* structure of your study, as well as to plan this study and carry it out. An essential feature of this model is that it treats research design as a real entity, not simply an abstraction or plan. Borrowing Kaplan's (1964, p. 8) distinction between the "logic-in-use" and "reconstructed logic" of research, this model can be used to represent the "design-in-use" of a study, the *actual* relationships among the components of the research, as well as the intended (or reconstructed) design (Maxwell & Loomis, 2002). The design of your research, like the design of the *Vasa,* is real and will have real consequences. As Yin stated, "every type of empirical research has an implicit, if not explicit, research design" (1994, p. 19). Because a design always exists, it is important to *make* it explicit, to get it out in the open where its strengths, limitations, and consequences can be clearly understood.

This conception of design as a model of, as well as for, research is exemplified in a classic qualitative study of medical students (Becker, Geer, Hughes, & Strauss, 1961). The authors began their chapter on the "design of the study" by stating that

> In one sense, our study had no design. That is, we had no well-worked-out set of hypotheses to be tested, no data-gathering instruments purposely designed to secure information relevant to these hypotheses, no set of analytic procedures specified in advance. Insofar as the term "design" implies these features of elaborate prior planning, our study had none.
>
> If we take the idea of design in a larger and looser sense, using it to identify those elements of order, system, and consistency our procedures did exhibit, our study had a design. We can say what this was by describing our original view of the problem, our theoretical and methodological commitments, and the way these affected our research and were affected by it as we proceeded. (1961, p. 17)

Thus, to design a study, particularly a qualitative study, you can't just develop (or borrow) a logical strategy in advance and then implement it faithfully. Design in qualitative research is an ongoing process that involves "tacking" back and forth between the different components of the design, assessing the implications of goals, theories, research questions, methods, and validity threats for one another.[2] It does not begin from a predetermined starting point or proceed through a fixed sequence of steps, but involves interconnection and interaction among the different design components. In addition, as Frank Lloyd Wright emphasized, the design of something must fit not only with its use, but also with its environment. You will need to continually assess how this design is actually working during the research, how it influences and is influenced by its environment, and to make adjustments and changes so that your study can accomplish what you want.

My model of research design, which I call an "interactive" model (I could just as well have called it "systemic"), has a definite structure. However, it is an interconnected and flexible structure. In this book, I describe the key components of a research design, and present a strategy for creating coherent and workable relationships among these components. I also provide (in Chapter 7) an explicit plan for moving from your design to a research proposal.

The model I present here has five components, which I characterize below in terms of the concerns that each is intended to address:

1. *Goals.* Why is your study worth doing? What issues do you want it to clarify, and what practices and policies do you want it to influence? Why do you want to conduct this study, and why should we care about the results?

2. *Conceptual Framework.* What do you think is going on with the issues, settings, or people you plan to study? What theories, beliefs, and prior research findings will guide or inform your research, and what literature, preliminary studies, and personal experiences will you draw on for understanding the people or issues you are studying?

3. *Research Questions.* What, specifically, do you want to understand by doing this study? What do you *not* know about the phenomena you are studying that you want to learn? What questions will your research attempt to answer, and how are these questions related to one another?

4. *Methods.* What will you actually do in conducting this study? What approaches and techniques will you use to collect and analyze your data? There are four parts of this component of your design: (1) the relationships that you establish with the participants in your study; (2) your selection of settings, participants, times and places of data collection, and other data sources such as documents (what is often called "sampling"); (3) your data collection methods; and (4) your data analysis strategies and techniques.

5. *Validity.* How might your results and conclusions be wrong? What are the plausible alternative interpretations and validity threats to these, and how will you deal with these? How can the data that you have, or that you could potentially collect, support or challenge your ideas about what's going on? Why should we believe your results?

These components are not substantially different from the ones presented in many other discussions of research design (e.g., LeCompte & Preissle, 1993; Miles & Huberman, 1994; Robson, 2002; Rudestam & Newton, 1992, p. 5). What is innovative is the way the relationships among the components are conceptualized. In this model, the different parts of a design form an integrated and interacting whole, with each component closely tied to several others, rather than being linked in a linear or cyclic sequence. The most important relationships among these five components are displayed in Figure 1.1.

There are also connections other than those emphasized here, some of which I have indicated by dashed lines. For example, if a goal of your study is to empower participants to conduct their own research on issues that matter to them, this will shape the methods you use, and, conversely, the methods that are feasible in your study will constrain your goals. Similarly, the theories and intellectual traditions you are drawing on in your research will have implications for what validity threats you see as most important and vice versa.

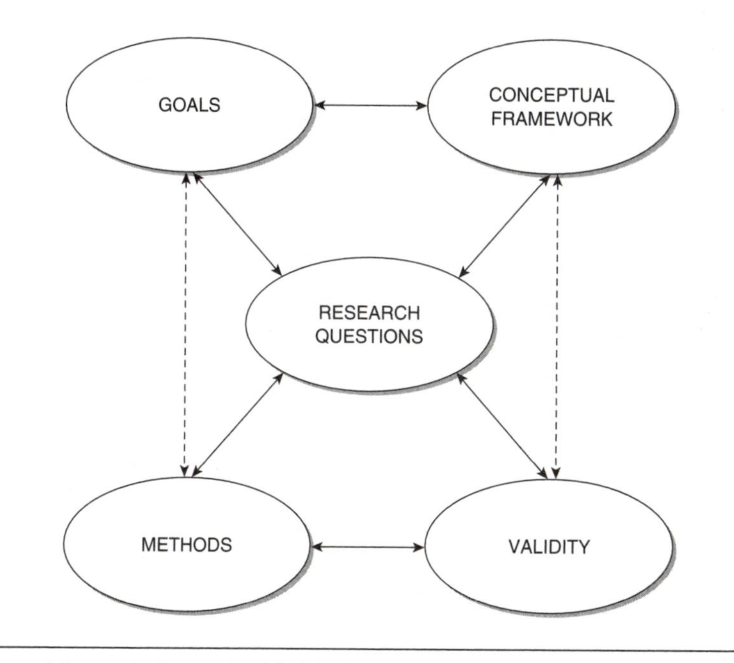

Figure 1.1 An Interactive Model of Research Design

The upper triangle of this model should be a closely integrated unit. Your research questions should have a clear relationship to the goals of your study, and should be informed by what is already known about the phenomena you are studying and the theoretical concepts and models that can be applied to these phenomena. In addition, the goals of your study should be informed by current theory and knowledge, while your decisions about what theory and knowledge are relevant depend on your goals and questions.

Similarly, the bottom triangle of the model should also be closely integrated. The methods you use must enable you to answer your research questions, and also to deal with plausible validity threats to these answers. The questions, in turn, need to be framed so as to take the feasibility of the methods and the seriousness of particular validity threats into account, while the plausibility and relevance of particular validity threats, and the ways these can be dealt with, depend on the questions and methods chosen. The research questions are the heart, or hub, of the model; they connect all of the other components of the design, and should inform, and be sensitive to, these components.

The connections among the different components of the model are not rigid rules or fixed implications; they allow for a certain amount of "give" and elasticity in the design. I find it useful to think of them as rubber bands. They can stretch and bend to some extent, but they exert a definite tension on different parts of the design, and beyond a particular point, or under certain stresses, they will break. This "rubber band" metaphor portrays a qualitative design as something with considerable flexibility, but in which there are constraints imposed by the different parts on one another, constraints which, if violated, make the design ineffective.

There are many other factors besides these five components that will influence the design of your study; these include your resources, research skills, perceived problems, ethical standards, the research setting, and the data you collect and results you draw from these data. In my view, these are not part of the *design* of a study, but either belong to the *environment* within which the research and its design exist or are *products* of the research. You will need to take these factors into account in designing your study, just as the design of a ship needs to take into account the kinds of winds and waves the ship will encounter and the sorts of cargo it will carry. Figure 1.2 presents some of the factors in the environment that can influence the design and conduct of a study, and displays

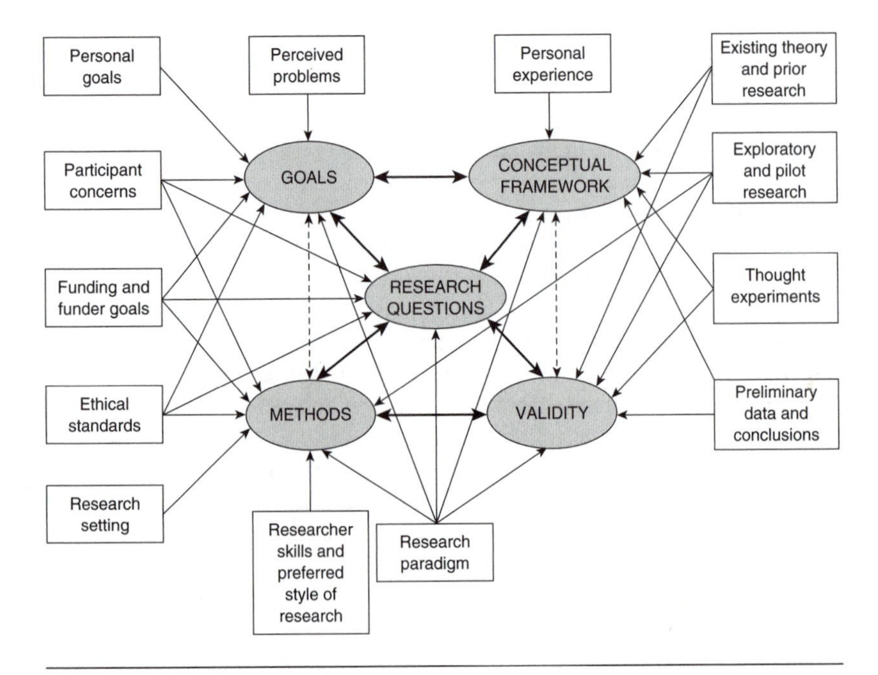

Figure 1.2 Contextual Factors Influencing a Research Design

some of the key linkages of these factors with components of the research design. These factors and linkages will be discussed in subsequent chapters.

I want to say something specifically about ethics, since I have not identified it as a separate component of research design. This isn't because I don't think ethics is important for qualitative design; on the contrary, attention to ethical issues in qualitative research is being increasingly recognized as essential (Christians, 2000; Denzin & Lincoln, 2000; Fine, Weis, Weseen, & Wong, 2000). Instead, it is because I believe that ethical concerns should be involved in *every* aspect of design. I have particularly tried to address these concerns in relation to methods, but they are also relevant to your goals, the selection of your research questions, validity concerns, and the critical assessment of your conceptual framework.

As the subtitle of this book indicates, my approach to design is an interactive one. It is interactive in three senses. First, the design model itself is interactive; each of the components has implications for the other components, rather than the components being in a linear, one-directional relationship with one another. Second, the design of a qualitative study should be able to change in response to the circumstances under which the study is being conducted, rather than simply being a fixed determinant of research practice. (Example 1.1 illustrates both of these interactive processes in the evolution of the design of one study.) Finally, the learning process embodied in this book is interactive, with frequent exercises that enable you to work on the design of your own study. This book does not simply present abstract research design principles that you can memorize and then later use in your research. You *will* learn generalizable principles, but you'll learn these best by creating a design for a particular qualitative project.

EXAMPLE 1.1

The Evolution of a Research Design

Maria Broderick began her dissertation study of a hospital-based support group for cancer patients with a theoretical background in adult psychological development and practical experience in the design of such programs; a research interest in discovering how patients' perceptions of support and interaction within the group were related to their developmental level; a plan to use observation, interviews, and developmental tests to answer this question; and the goals of improving such programs and developing a career in clinical practice. However, after her proposal was approved, she lost access to the group she had originally planned to study, and was unable to find another suitable cancer program. She ended up negotiating permission to study a stress-reduction program for patients in a hospital setting, but was not allowed to observe the classes; in addition, the program team insisted on a quasi-experimental research design, with pre- and postintervention measures of patients' developmental level and experiences. This forced her both to broaden her theoretical framework beyond cancer support programs to behavioral medicine programs in general and to alter her methods to rely primarily on pre- and postinterviews and developmental tests.

As Maria was beginning her research, she herself was diagnosed with a stress-related illness. This had a profound effect on the research design. First, she gained access to the program as a patient, and discovered that it wasn't actually run as a support program, but in a traditional classroom format. This made her extensive literature review on support groups largely irrelevant. Second, she found that her own experiences of her illness and what seemed to help her deal with stress differed substantially from what was reported in the literature. These two developments profoundly altered her conceptual framework and research questions, shifting her theoretical focus from ego development to cognitive development, adult learning, and educational theory. In addition, she found that pretesting of the patients was impossible for practical reasons, eliminating the possibility of quasi-experimental assessment of patient changes and shifting her methods and validity checks back toward her original plans.

While Maria was analyzing her data, her gradual creation of a theory that made sense of these patients' (and her own) experiences directed her to new bodies of literature and theoretical approaches. Her increasing focus on what the patients *learned* through the program caused her to see meditation and cognitive restructuring as tools for reshaping one's view of stress, and led her to develop a broader view of stress as a cultural phenomenon. It also reconnected her with her longtime interest in nontraditional education for adults. Finally, these changes led to a shift in her career goals from clinical practice to an academic position, and her goals for the study came to emphasize relating adult developmental theory to empowerment curricula and improving adult education in nontraditional settings.

One way in which the design model presented here can be useful is as a tool or template for conceptually mapping the design of an actual study, either as part of the design process or in analyzing the design of a completed study. This involves filling in the circles for the five components of the model with the

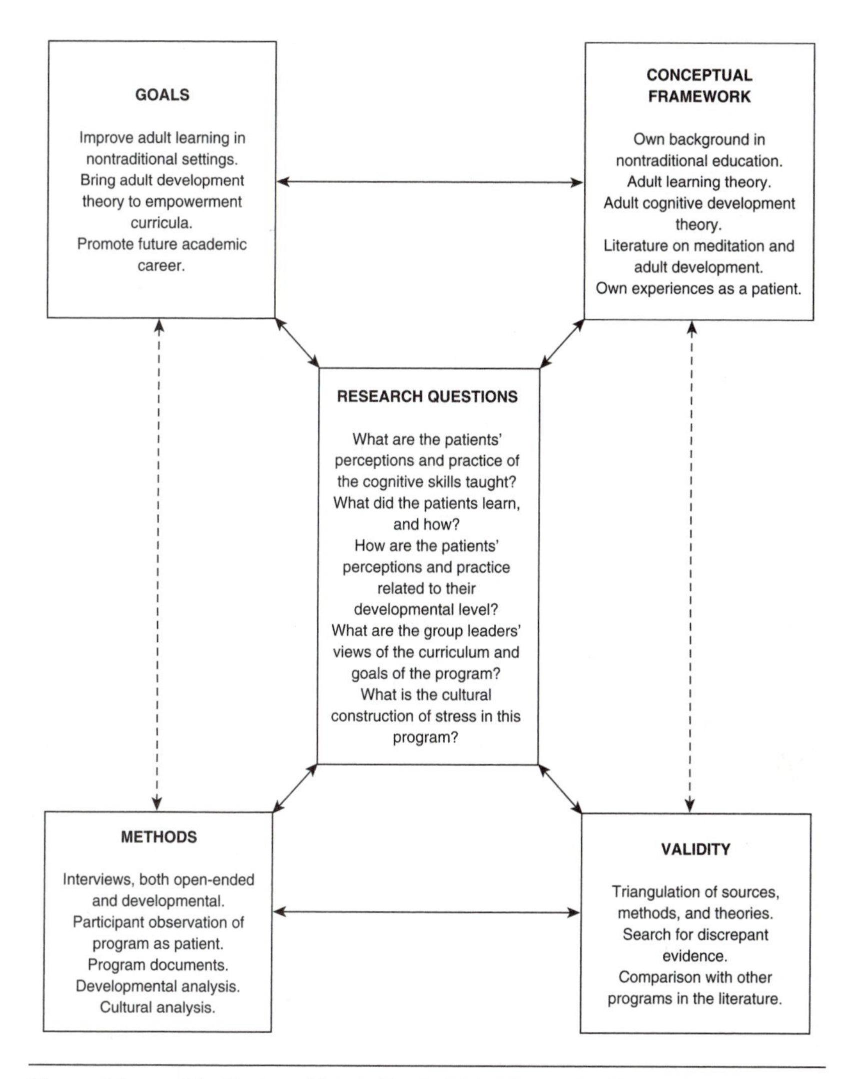

Figure 1.3 The Design of Maria Broderick's Dissertation Research

specific components of that study's design, a strategy that I call a "design map." Figure 1.3 is a design map of the eventual structure of Maria Broderick's dissertation research; see Maxwell and Loomis (2002) for other such maps.

I do not believe that there is one right model of, or for, research design. However, I think that the model that I present here is a useful model for two main reasons:

1. It explicitly identifies as *components* of design the key issues about which you will need to make decisions and which will need to be addressed in your research proposal. These components are therefore less likely to be overlooked or misunderstood, and can be dealt with in a systematic manner.
2. It emphasizes the *interactive* nature of design decisions in qualitative research and the multiple connections among design components. A common reason that dissertation or funding proposals are rejected is because they do not make clear the logical connections among the design components, and the model I present here makes it easier to understand and demonstrate these connections.

A good design for your study, like a good design for a ship, will help it to safely and efficiently reach its destination. A poor design, one in which the components are not well integrated or are incompatible, will at best be inefficient, and at worst will fail to achieve its goals.

THE ORGANIZATION OF THIS BOOK

This book is structured to guide you through the process of designing a qualitative study. It highlights the issues for which design decisions must be made, and presents some of the considerations that should inform these decisions. Each chapter in the book deals with one component of design, and these chapters form a logical sequence. However, this organization is only a conceptual and presentational device, not a procedure to follow in designing an actual study. You should make decisions about each component in light of your thinking about all of the other components, and you may need to modify previous design decisions in response to new information or changes in your thinking.

This book takes a Z-shaped path (Figure 1.4) through the components of this model, beginning with goals (Chapter 2). The goals of your study are not only important, but also primary; if your reasons for doing the study aren't clear, it can be difficult to make *any* decisions about the rest of the design. Your conceptual framework (Chapter 3) is discussed next, both because it should connect closely to your goals and because the goals and framework jointly have a major influence on the formulation of research questions for the study. Your research questions (Chapter 4) are thus a logical next topic; these three components should form a coherent unit.

The next component discussed is methods (Chapter 5): how you will actually collect and analyze the data to answer your research questions. However, these methods and analyses need to be connected to issues of validity (Chapter 6): how you might be wrong, and what would make your answers more believable than alternative possible answers. Research questions, methods, and validity also

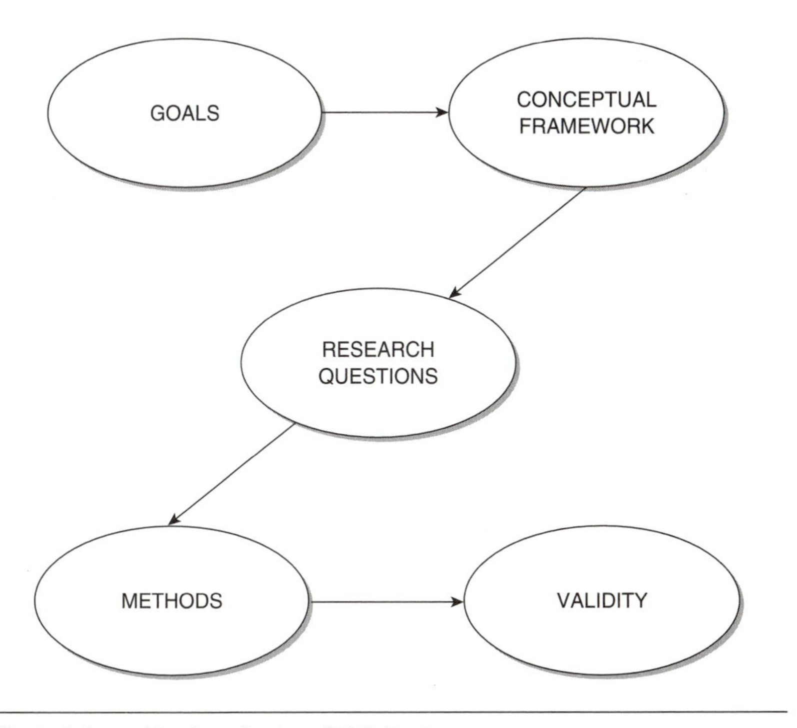

Figure 1.4 The Organization of This Book

should form an integrated unit, one in which the methods for obtaining answers to the questions, and the means for assuring the credibility of the potential answers in the face of plausible validity threats, are clearly conceptualized and linked to the research questions.

Finally, Chapter 7 discusses the implications of my model of design for developing research proposals, and provides a map and guidelines for how to get from one to the other.

THE EXERCISES IN THIS BOOK

C. Wright Mills wrote that

> One of the very worst things that happens to social scientists is that they feel the need to write of their "plans" on only one occasion: when they are going to ask for money for a specific piece of work or "a project." It is as a request for funds

> that most planning is done, or at least carefully written about. However standard the practice, I think this very bad: it is bound in some degree to be salesmanship, and, given prevailing expectations, very likely to result in painstaking pretensions; the project is likely to be "presented," rounded out in some manner long before it ought to be; it is often a contrived thing, aimed at getting the money for ulterior purposes, however valuable, as well as for the research presented. A practicing social scientist ought periodically to review "the state of my problems and plans." (Mills, 1959, p. 197).

He went on to make an eloquent plea that each researcher write regularly and systematically about his or her research, "just for himself and perhaps for discussion with friends" (p. 198), and to keep a file of these writings, which qualitative researchers usually call "memos."

All of the exercises in this book are memos of one sort or another, and I want to briefly discuss the nature of memos and how to use them effectively. Memos (sometimes called "analytic memos") are an extremely versatile tool that can be used for many different purposes. This term refers to any writing that a researcher does in relationship to the research other than actual field notes, transcription, or coding. A memo can range from a brief marginal comment on a transcript or a theoretical idea recorded in a field journal to a full-fledged analytic essay. What all of these have in common is that they are ways of getting ideas down on paper (or in a computer), and of using this writing as a way to facilitate reflection and analytic insight. When your thoughts are recorded in memos, you can code and file them just as you do your field notes and interview transcripts, and return to them to develop the ideas further. Not writing memos is the research equivalent of having Alzheimer's disease; you may not remember your important insights when you need them. Peters (1992, p. 123) cited Lewis Carroll's *Through the Looking Glass* on this function of memos:

> "The horror of that moment," the King went on, "I shall never, *never* forget."
>
> "You will, though," said the Queen, "unless you make a memorandum of it."

Many of the examples used in this book are memos, or are based on memos.[3]

Memos are one of the most important techniques you have for developing your own ideas. You should therefore think of memos as a way to help you *understand* your topic, setting, or study, not just as a way of recording or presenting an understanding you've already reached. Memos should include reflections on your reading and ideas as well as your fieldwork. Memos can be written on methodological issues, ethics, personal reactions, or anything else; I wrote numerous memos about research design during the writing and revising of this book. Write memos as a way of working on a problem you encounter in making sense of your topic, setting, study, or data. Write memos whenever you have an idea that you want to develop further, or simply to record the idea for later development. Write *lots* of memos throughout the course of your research project; remember that in qualitative research, design is something that goes on during the entire study, not just at the beginning. Think of memos as a kind of decentralized field journal; if you prefer, you can write your memos in an actual journal.

Whatever form these memos take, their value depends on two things. The first is that you engage in serious reflection, analysis, and self-critique, rather than just mechanically recording events and thoughts. The second is that you *organize* your memos in a systematic, retrievable form, so that the observations and insights can easily be accessed for future examination. I do my own memo writing primarily in two forms: on 3 × 5 cards, which I always carry with me for jotting down ideas and which I index by date and topic, and in computer files relating to particular projects, which I use for longer memos. During my dissertation research in an Inuit community in northern Canada, I also kept a field journal, which was invaluable in making sense of my personal responses to the research situation. It can also be very useful to share some of your memos with colleagues or fellow students for their feedback.[4]

Although memos are primarily a tool for thinking, they can also serve as an initial draft of material that you will later incorporate (usually with substantial revision) in a proposal, report, or publication, and I've tried to design most of the memo exercises in this book so that they can be used in this way. However, thinking of memos primarily as a way of communicating to *other* people will usually interfere with the kind of reflective writing that you need to do to make memos most useful to you. In particular, beware of what Becker (1986) called "classy writing"—pretentious and verbose language that is intended to impress others rather than to clarify your ideas. A saying among writing instructors is, "When you write, don't put a tuxedo on your brain" (Metzger, 1993).

NOTES

1. A third definition treats designs as distinct, standard arrangements of research methods that have their own coherence and logic, as possible answers to the question, "What research design are you using?" For example, a randomized, double-blind experiment is one research design; a qualitative case study is another. For several reasons, this typological conception of design is not particularly helpful in qualitative research. First, few qualitative studies lend themselves to such "off-the-rack" approaches; as discussed throughout this book, qualitative design needs to be inductive, flexible, and tailored to the specific situation being studied. Second, typological approaches to

design generally deal explicitly only with methods, and neither address the other components of design (in my model, goals, conceptual frameworks, research questions, and validity) nor clarify the actual functioning and interrelationship of the parts of a design. For a more detailed analysis of the strengths and limitations of typological approaches to design, see Maxwell and Loomis (2002).

2. This tacking back and forth is similar in some ways to the "hermeneutic circle" of textual interpretation (Geertz, 1974). However, I am advocating an interactive rather than a sequential model of research design primarily because I see design as pertaining to the actual relationships of the components of a research study, not because I take an "interpretive" or "humanistic" as opposed to a "scientific" view of research. The interactive model I present here is drawn to a significant extent from research practices in the natural sciences, particularly biology, and is applicable to quantitative as well as qualitative research (Maxwell & Loomis, 2002). In contrast, Janesick (1994), who saw qualitative research design as an interpretive art form analogous to dance, nevertheless stated that "qualitative research design begins with a question" (p. 210) and presented research design as a sequence of decisions that the researcher will need to make at each stage of the research.

3. For additional discussion and examples of what a memo involves, see Bogdan and Biklen (2003, pp. 114–116, 151–157), Miles and Huberman (1994, pp. 72–75), and Mills (1959). More detailed information on memos can be found in Strauss (1987, chaps. 1, 5, and 6) and Strauss and Corbin (1990, chap. 12).

4. See Mills (1959) for advice on how to use memos in developing a research agenda and career.

2

Goals

Why Are You Doing This Study?

In planning, as well as in assessing, ethnographic research, we must consider its relevance as well as its validity.

—Hammersley, 1992, p. 85

Anyone can find an unanswered, empirically answerable question for which the answer isn't *worth* knowing; as Thoreau said, it is not worthwhile to go around the world to count the cats in Zanzibar. In addition, it is easy to become captivated by the stories of your informants, or by what's going on in the setting you are studying, and lose sight of your *reasons* for studying these particular phenomena. Brendan Croskery (1995), reflecting on his dissertation research on four Newfoundland school principals, admitted that

> The study suffered from too many good intentions and too little focused thinking. . . . I painfully discovered that many of the data (though interesting) were not particularly relevant to the core category. (p. 348)

A clear understanding of the goals motivating your work will help you to avoid losing your way or spending time and effort doing things that don't advance these goals.

The goals of your study are an important part of your research design. (I am using "goal" in a broad sense to include motives, desires, and purposes—anything that leads you to do the study or that you want to accomplish by doing it.[1]) These goals serve two main functions for your research. First, they help to guide your other design decisions to ensure that your study is *worth* doing, that you get something of value out of it. Second, they are essential to *justifying* your study, a key task of a funding or dissertation proposal. In addition, as Hammersley (1992, p. 28) noted, your goals inevitably shape the descriptions, interpretations, and theories you create in your research. They therefore constitute not only important *resources* that you can draw on in planning, conducting, and justifying the research, but also potential *validity threats,* or sources of bias for the research results, that you will need to deal with (see Chapter 6).

PERSONAL, PRACTICAL, AND INTELLECTUAL GOALS

It is useful to distinguish among three different kinds of goals for doing a study: personal goals, practical goals, and intellectual (or scholarly) goals. Personal goals are things that motivate *you* to do the study, but are not necessarily important for others. They can include the desire to change or improve some situation that you're involved in, curiosity about a specific topic or event, a preference for conducting a particular type of research, or simply the need to advance your career. These personal goals often overlap with your practical or research goals, but they may also include deeply rooted individual desires and needs that bear little relationship to your "official" reasons for doing the study (see Example 2.1).

Two major decisions are often profoundly influenced by the researcher's personal goals. One is the topic, issue, or question selected for study. Traditionally, students have been told to base this decision on either faculty advice or the literature on their topic. However, personal goals and experiences play an important role in many research studies. Strauss and Corbin (1990) argued that

> choosing a research problem through the professional or personal experience route may seem more hazardous than through the suggested [by faculty] or literature routes. This is not necessarily true. The touchstone of your own experience may be more valuable an indicator for you of a potentially successful research endeavor. (pp. 35–36)

A particularly important advantage of basing your research topic on your own experience is *motivation.* Lack of motivation causes many students to never finish their dissertations, and a strong personal interest in the topic and in answering your research questions can counteract the inevitable interference from work, family obligations, or just procrastination. Example 2.1 describes how one student made a substantial change in her dissertation topic as a result of her own life experiences and the goals and interests that these created.

EXAMPLE 2.1

Using Personal Experience to Choose a Dissertation Topic

Carol Kaffenberger, a doctoral student in a counseling program, had carefully planned her dissertation research on the development of conflict resolution skills in children, and was beginning work on her dissertation proposal. However, she found it hard to sustain her interest in this topic. Three years before she began her doctoral work, her youngest daughter, then 12, had been diagnosed with a particularly deadly form of leukemia, was hospitalized for 6 months and underwent a bone marrow transplant, went into remission and then relapsed, and required a second transplant before recovering 3 years later. This illness had initiated a family crisis, and caused major changes in the family's roles and responsibilities. Carol quit her job and moved into the hospital with her daughter. Her husband continued to work, maintained the house, and parented their son, who was 15 at the time of the diagnosis. Their older daughter was away at college, but was the donor for the bone marrow transplants.

Initially, Carol had felt that her family was coping well, but as the crisis wore on, she was surprised by the amount of anger and emotional distress expressed by the older siblings, anger that, despite her counseling training, she did not understand. Watching her family getting "back to normal" after this ordeal, she realized they were never going to be the same. She also realized that her prior assumptions about their experience had been incorrect, and she became very interested in understanding this experience.

At a doctoral student meeting, another student, who knew of Carol's involvement with her daughter's cancer, asked her about her dissertation plans. Carol replied that she would be looking at children's development of conflict resolution skills, and briefly described her plans. The student replied, "What a missed opportunity!" explaining that she thought studying the consequences for families of adolescent cancer would be a terrific topic. After thinking about this, Carol went to her advisor, mentioned the student's idea, and asked, "Is this crazy?" Her advisor replied, "I've been waiting for you to be ready to do this."

Carol did a literature review and found that little was known about the meaning and consequences of adolescent cancer for families, particularly for siblings. She also found that, with increasing survival rates, schools were dealing with many more students who had been affected by a lengthy experience with cancer, as either a survivor or the sibling of a survivor, but had little experience in handling these issues. Motivated by her own interest in this topic, the lack of available information, and the growing importance of this issue, she changed her dissertation to a study of the long-term impact and meaning of adolescent cancer for survivors and their siblings, and its effect on the sibling relationship. She enrolled in my dissertation proposal course in the fall of 1997, defended her proposal in the spring of 1998, and defended her dissertation 1 year later. She says that she "loved every minute of her dissertation"; she even took her data with her on a vacation to Bermuda when she was finishing her data analysis (Kaffenberger, 1999, personal communication).

A second decision that is often influenced by personal goals and experiences is the choice of a qualitative approach. Locke, Spirduso, and Silverman (1993) argued that "every graduate student who is tempted to employ a qualitative design should confront one question, 'Why do I want to do a qualitative study?' and then answer it honestly" (p. 107). They emphasized that qualitative research is *not* easier than quantitative research and that seeking to avoid statistics bears little relationship to having the personal interests and skills that qualitative inquiry requires (pp. 107–110). The key issue is the compatibility of your reasons for "going qualitative" with your other goals, your research questions, and the actual activities involved in doing a qualitative study. Alan Peshkin's motives (Example 2.2) for doing qualitative research—that he liked qualitative fieldwork and that it suited his abilities—are perfectly legitimate ones, *if* you choose research questions for which this is an appropriate strategy.

Traditionally, discussions of personal goals in research methods texts have accepted, implicitly or explicitly, the ideal of the objective, disinterested scientist, and have emphasized that the choice of research approaches and methods should be determined by the research questions that you want to answer. However, it is clear from autobiographies of scientists (e.g., Heinrich, 1984) that decisions about research methods are often far more personal than this, and the importance of subjective motives and goals in science is supported by a great deal of historical, sociological, and philosophical work.

The grain of truth in the traditional view is that your personal (and often unexamined) motives as researcher have important consequences for the validity of your conclusions. If your data collection and analysis are based on personal desires *without* a careful assessment of the implications of the latter for your methods and conclusions, you are in danger of creating a flawed or biased study.

King Gustav of Sweden wanted a powerful warship to dominate the Baltic, but this desire led to an ill-considered decision to add a second gundeck to the *Vasa,* causing it to capsize and sink and thus dealing a severe setback to his goals.

For all of these reasons, it is important that you recognize and take account of the personal goals that drive and influence your research. Attempting to exclude your personal goals and concerns from the design of your research is neither possible nor necessary. What *is* necessary is to be *aware* of these goals and how they may be shaping your research, and to think about how best to achieve them *and* to deal with their influence. In addition, recognizing your personal ties to the study you want to conduct can provide you with a valuable source of insight, theory, and data about the phenomena you are studying (Marshall & Rossman, 1999, pp. 25–30; Strauss & Corbin, 1990, pp. 42–43); this source will be discussed in the next chapter in the section titled "Experiential Knowledge." Example 2.2 describes how one researcher's personal goals and values influenced (and were influenced by) a series of qualitative studies.

EXAMPLE 2.2

The Importance of Personal Values and Identity

Alan Peshkin's personal goals, rooted in his own values and identity, profoundly influenced several ethnographic studies he did of schools and their communities (1991, pp. 285–295; Glesne & Peshkin, 1992, pp. 93–107). In his first study, in a rural town he called Mansfield, he liked the community and felt protective toward it. This shaped the kind of story that he told, a story about the importance of community and its preservation. In contrast, in his second study, an ethnography of a fundamentalist Christian school (which he called Bethany Baptist Academy, BBA) and its community, he felt alienated, as a Jew, from a community that attempted to proselytize him:

> When I began to write . . . I knew I was annoyed by my *personal* (as opposed to research) experience at BBA. I soon became sharply aware that my annoyance was pervasively present, that I was writing out of pique and vexation. Accordingly, I was not celebrating community at Bethany, and community prevailed there no less robustly than it had at Mansfield. Why not? I was more than annoyed in Bethany; my ox had been gored. The consequence was that the story I was feeling drawn to tell had its origins in my personal sense of threat. I was not at Bethany as a cool, dispassionate observer (are there any?); I was there as a Jew whose otherness was

> dramatized directly and indirectly during eighteen months of fieldwork. (Glesne & Peshkin, 1992, p. 103)

In hindsight, Peshkin realized that if he had been less sympathetic toward Mansfield, he could have told a different, equally valid story about this community, whereas if he had identified with Bethany and wanted to support and perpetuate it, he could legitimately have showed how it was much like Mansfield.

In a third study, this one of an urban, multiethnic and multiracial school and community that he called Riverview, Peshkin resolved at the outset to try to identify the aspects of his identity that he saw emerging in his reactions. He listed six different subjective "I's" that influenced this study, each embodying its own goals. These included the Ethnic-Maintenance I and the Community-Maintenance I that he had discovered in his earlier studies; an E-Pluribus-Unum I that supported the ethnic and racial "mingling" that he saw going on; a Justice-Seeking I that wanted to correct the negative and biased images of Riverview held by its wealthier neighbors; a Pedagogical-Meliorist I that was disturbed by the poor teaching that many minority students received in Riverview and sought to find ways to improve this; and a Nonresearch-Human I that was grateful for the warm reception he and his wife received in Riverview, generated a concern for the people and community, and moderated otherwise sharp judgments he might have made.

Peshkin strongly recommended that all researchers systematically monitor their subjectivity:

> I see this monitoring as a necessary exercise, a workout, a tuning up of my subjectivity to get it in shape. It is a rehearsal for keeping the lines of my subjectivity open—and straight. And it is a warning to myself so that I may avoid the trap of perceiving just what my own untamed sentiments have sought out and served up as data. (Peshkin, 1991, pp. 293–294)

Exercise 2.1 is one way to engage in this monitoring.

In addition to influencing his questions and conclusions, Peshkin's personal goals were intimately involved in his choice of methods. As he stated, "I like fieldwork, it suits me, and I concluded that rather than pursuing research with questions in search of the 'right' methods of data collection, I had a preferred method of data collection in search of the 'right' question" (Glesne & Peshkin, 1992, p. 102).

In addition to your personal goals, there are two other kinds of goals (ones that are important for other people, not just yourself) that I want to distinguish and discuss. These are practical goals (including administrative or policy goals) and intellectual goals. Practical goals are focused on *accomplishing* something—meeting some need, changing some situation, or achieving some objective. Intellectual goals, in contrast, are focused on *understanding* something—gaining insight into what is going on and why this is happening, or answering some question that previous research has not adequately addressed.

Both of these kinds of goals are legitimate parts of your design. However, they need to be distinguished, because while intellectual goals are often a fruitful starting point for framing research questions, practical goals can't normally be used in this straightforward way. Research questions need to be questions that your study can potentially answer, and questions that ask directly about how to accomplish practical goals, such as "How should this program be modified to make it more equitable?" or "What can be done to increase students' motivation to learn science?" are not directly answerable by any research. Such questions have an inherently open-ended nature (expressed by terms such as "can") or value component (expressed by terms such as "should") that no amount of data or analysis can fully address.

On the other hand, research questions such as "What effect has this new policy had on program equity?" or "How did students respond to this new science curriculum?" are not only potentially answerable, but can advance the practical goals implied in the previous questions. For these reasons, you need to frame your research questions in ways that help your study to *achieve* your practical goals, rather than smuggling these goals into the research questions themselves, where they may interfere with the coherence and feasibility of your design. A common problem that my students have in developing research questions is that they try to base these questions directly on their practical goals, ending up with questions that not only can't be answered by their research, but fail to adequately guide the research itself. I will discuss this issue more fully in Chapter 4; here, I am simply emphasizing the difference between these two types of goals.

The point is not to eliminate practical goals from your design; in addition to the reasons given previously, practical or policy objectives are particularly important for *justifying* your research. Don't ignore these goals, but understand where they are coming from, their implications for your research, and how they can be productively employed in planning and defending your study.

WHAT GOALS CAN QUALITATIVE RESEARCH HELP YOU ACHIEVE?

Qualitative and quantitative methods are not simply different ways of doing the same thing. Instead, they have different strengths and logics, and are often best used to address different kinds of questions and goals (Maxwell & Loomis, 2002). The strengths of qualitative research derive primarily from its inductive approach, its focus on specific situations or people, and its emphasis on words rather than numbers.

I will describe five particular *intellectual* goals for which qualitative studies are especially suited, and three *practical* goals to which these intellectual goals can substantially contribute:

1. Understanding the *meaning,* for participants in the study, of the events, situations, experiences, and actions they are involved with or engage in. I am using "meaning" here in a broad sense, including cognition, affect, intentions, and anything else that can be encompassed in what qualitative researchers often refer to as the "participants' perspective." This perspective is not simply their account of these events and actions, to be assessed in terms of its truth or falsity; it is *part of* the reality that you are trying to understand (Maxwell, 1992; Menzel, 1978). In a qualitative study, you are interested not only in the physical events and behavior that are taking place, but also in how the participants in your study make sense of these, and how their understanding influences their behavior. This focus on meaning is central to what is known as the "interpretive" approach to social science (Bredo & Feinberg, 1982; Geertz, 1974; Rabinow & Sullivan, 1979).

2. Understanding the particular *context* within which the participants act, and the influence that this context has on their actions. Qualitative researchers typically study a relatively small number of individuals or situations, and preserve the individuality of each of these in their analyses, rather than collecting data from large samples and aggregating the data across individuals or situations. Thus, they are able to understand how events, actions, and meanings are shaped by the unique circumstances in which these occur (Maxwell, 2004a).

3. Identifying *unanticipated* phenomena and influences, and generating new, "grounded" theories about the latter. Qualitative research has an inherent openness and flexibility that allows you to modify your design and focus during the research to understand new discoveries and relationships. This flexibility derives from its particularistic, rather than comparative and generalizing, focus, and from its freedom from the rules of statistical hypothesis testing, which require that the research plan not be significantly altered after data collection has begun.

4. Understanding the *process* by which events and actions take place. Merriam stated that "The interest [in a qualitative study] is in process rather than outcomes" (1988, p. xii); while this does not mean that qualitative research is unconcerned with outcomes, it does emphasize that a major strength of qualitative research is in getting at the processes that led to these outcomes, processes that experimental and survey research are often poor at identifying (Britan, 1978; Maxwell, 2004a, 2004c; Patton, 1990, p. 94).

5. Developing *causal explanations*. The traditional view that only quantitative methods can be used to credibly draw causal conclusions has long been disputed by some qualitative researchers (e.g., Britan, 1978; Denzin, 1970; Erickson, 1986). Miles and Huberman (1984) argued that

> Much recent research supports a claim that we wish to make here: that field research is far *better* than solely quantified approaches at developing explanations of what we call local causality—the actual events and processes that led to specific outcomes. (p. 132)

Although the traditional view has been abandoned by some researchers, both qualitative and quantitative (e.g., Shadish, Cook, & Campbell, 2002; cf. Maxwell, 2004a, 2004c), it is still dominant in both traditions (Denzin & Lincoln, 2000; Shavelson & Towne, 2002).

Part of the reason for the disagreement has been a failure to recognize that quantitative and qualitative researchers tend to ask different kinds of causal questions. Quantitative researchers tend to be interested in whether and to what extent *variance* in *x* causes variance in *y*. Qualitative researchers, on the other hand, tend to ask *how x* plays a role in causing *y*, what the *process* is that connects *x* and *y*. Mohr (1982) used the terms "variance theory" and "process theory" to refer to these two general approaches to research, and I will return to this distinction in later chapters. This emphasis on understanding processes and mechanisms, rather than demonstrating regularities in the relationships between variables, is fundamental to realist views of causation, which are prominent in the current philosophy of science (Maxwell, 2004a). Weiss (1994) provided a concrete illustration of this difference:

> In qualitative interview studies the demonstration of causation rests heavily on the description of a visualizable sequence of events, each event flowing into the next. . . . Quantitative studies support an assertion of causation by showing a correlation between an earlier event and a subsequent event. An analysis of data collected in a large-scale sample survey might, for example, show that there is a correlation between the level of the wife's education and the presence of a companionable marriage. In qualitative studies we would look for a process

> through which the wife's education or factors associated with her education express themselves in marital interaction. (p. 179)

This is not to say that deriving causal explanations from a qualitative study is an easy or straightforward task (Maxwell, 2004c). However, the situation of qualitative research is no different from that of quantitative research in this respect. Both approaches need to identify and deal with the plausible validity threats to any proposed causal explanation; I will discuss this further in Chapter 6.

These intellectual goals, and the inductive, open-ended strategy that they require, give qualitative research a particular advantage in addressing three practical goals:

6. Generating results and theories that are understandable and experientially credible, both to the people you are studying and to others. Patton (1990, pp. 19–24) gave an example of how the responses to the open-ended items on a questionnaire used to evaluate a teacher accountability system had far greater credibility with, and impact on, the school administration than did the quantitative analysis of the standardized items. Bolster (1983) made a more general argument—namely, that one of the reasons for the lack of impact of educational research on educational practice has been that such research has largely been quantitative, and doesn't connect with teachers' experience of everyday classroom realities. He argued for a qualitative approach that emphasizes the perspective of teachers and the understanding of particular settings, as having far more potential for informing educational practitioners.

7. Conducting formative evaluations, ones that are intended to help improve existing practice rather than to simply assess the value of the program or product being evaluated (Scriven, 1967, 1991). In such evaluations, it is more important to understand the process by which things happen in a particular situation than to rigorously compare this situation with others.

8. Engaging in collaborative or action research with practitioners or research participants. The "face credibility" of qualitative research, and its focus on particular contexts and their meaning for the participants in these contexts, make it particularly suitable for collaborations with these participants (Reason, 1988, 1994; Tolman & Brydon-Miller, 2001). In addition, there are important ethical arguments for incorporating the perspectives and goals of these participants in your research design (Lincoln, 1990).

Sorting out and assessing the different personal, practical, and intellectual goals that you bring to your study can be a difficult task. In addition, this is not something you can do once, when you begin designing the study, and then forget about, as Example 2.2 illustrates. Some of your goals may not become apparent to you until you are well into the research; furthermore, they may change as the research proceeds. Example 2.3 provides an account of how one doctoral student went about identifying her goals in making a decision about her dissertation topic. Exercise 2.1, at the end of this chapter, is what I call a "researcher identity memo"; it asks you to write about the goals and personal identity that you bring to your study, and their potential benefits and liabilities for your research. Example 2.4 is one such memo, written for my qualitative methods class; it shows how one student wrestled with deep and painful issues of her own identity and goals in planning for her dissertation research on language curriculum reform in Bolivia. All of the examples in this chapter illustrate some of the advantages that reflection on your goals can provide for your research.

EXAMPLE 2.3

Deciding on a Dissertation Topic

During her first year of doctoral work, Isabel Londoño, a native of Colombia, enrolled in a qualitative research methods course. For her research project, she interviewed seven women from her country who were working in Boston, exploring their experiences balancing work and family. While working on the project, she also began to read some of the feminist literature available in the United States on women executives, women's psychological development, and women's experience managing work and family. She was excited by the new ideas in this literature, which she had not had access to in her own country, and decided that she wanted to focus on issues of executive women in her country for her dissertation.

At the end of her first year, Isabel took a leave of absence from the doctoral program to work as the chief of staff of her former college roommate, whose husband had just been elected president of Colombia. Among her responsibilities was gathering information on employment, education, and the status of women in her nation. One of the issues that emerged as critical was the need to assess the effect of a recent shift in educational decision making from the national to the local level. In the past, most decisions had been made by the national ministry of education; now decisions were being shifted downward to the mayors in local municipalities. No one was really sure how this change was being implemented and what its effects were.

Isabel found that investigating an issue that affected the lives of many people in her country changed her perspective, and raised questions about her choice of a thesis topic:

> It became an issue of what was my responsibility to the world. To find out how to solve a personal, internal conflict of executive women? Or was there a problem where I could really be of help? Also, what was more rewarding to me as a person—to solve a problem that affected me personally or solve a problem of the world?

She also felt pressure from others to select a topic that clearly linked to her career goals and showed that she knew what she wanted to do with her life.

Coming to a decision about her dissertation research topic forced Isabel to identify and assess her personal and practical goals.

> I thought about why I got into a doctoral program. What I hoped to get out of it personally, professionally, academically. Why did I end up here? Then, I thought about what are the things about the world that move me, that make me sad or happy? I analyzed what that interest was about—people, feelings, institutions. It was important for me to see the themes in common in my interests and motivations. It gave me strength. I also was open to change. Change is the most scary thing, but you have to allow it.

She decided that she would study the decentralization of educational decision making in six municipalities in her country. In making this decision, she chose to disregard others' opinions of her:

> What I have decided is *no,* I am going to do my thesis about something that *moves me inside.* I don't care if I am ever going to work on that topic again because it's something I want to learn about. I don't want to use my thesis as a stepladder for my work, that feels like prostitution. So I believe the interest should be on the thesis topic itself, not on where that is leading you, where you're going to get with it.

One of the things that supported her decision was reading the literature on her topic:

> That was very important because I discovered that what I was interested in was something that had interested a lot of other people before, and was going on in a lot of other places in the world, and was affecting education in other countries. This made my topic relevant. It was very important for

me to understand that it was relevant, that I was not just making up a dream problem. I think that's something you always fear, that the problem you see is not really important. I also learned that although other people had done work on the problem, *nobody* had the interest I had—the human impact of implementing a reform in the administration of education.

Writing memos for classes was key, having to put things to paper. I also started keeping a thesis diary and wrote memos to myself in it. The date and one word, one idea, or something that I'd read. Many of the things I've written about have now become the list of what I'm going to do *after* I do my thesis!

Finally, I think it's important to really try to have fun. I figure, if you don't have fun, you shouldn't be doing it. Of course, sometimes I get tired of my topic and hate it. I sit at the computer and I'm tired and I don't want to do it, but every time I start working, I forget all that and get immersed in my work. And if something has the power to do that, it must be right.

The particular decisions that Isabel made are not necessarily the right ones for everyone; they are unique to her own identity and situation. However, the *way* that she went about making the decision—seriously and systematically reflecting on her goals and motives, and the implications of these for her research choices—is one that I recommend to everyone deciding on a major research project.

EXERCISE 2.1

Researcher Identity Memo

The purpose of this memo is to help you examine your goals, experiences, assumptions, feelings, and values as they relate to your research, and to discover what resources and potential concerns your identity and experience may create. What prior connections (social and intellectual) do you have to the topics, people, or settings you plan to study? How do you think and feel about these topics, people, or settings? What assumptions are you making, consciously or unconsciously, about these? What do you want to accomplish or learn by doing this study?

The purpose of this exercise is not to write a *general* account of your goals, background, and experiences. Instead, identify those goals and experiences, and the beliefs and emotions that connect to these, that are most relevant to your planned research, and reflect on *how* these have informed and influenced your research. See Examples 2.2, 2.3, and 2.4 for some of the things you can do with such a memo—not as *models* to mechanically follow, but as *illustrations* of the kind of thinking that this memo requires. If you are just starting your project, you can't be as detailed or confident in your conclusions as some of these researchers are, but try to aim for this sort of exploration of how your identity and goals could affect your study.

The memo is intended to be mainly for *your* benefit, not for communicating to someone else; try to avoid substituting presentation for reflection and analysis. I suggest that you begin working on this memo by "brainstorming" whatever comes to mind when you think about your prior experiences that may relate to your site or topic, and jot these down without immediately trying to organize or analyze them. Then, try to identify the issues most likely to be important in your research, think about the implications of these, and organize your reflections.

Below are two broad sets of questions that it is productive to reflect on in this memo. In your answers to these, try to be as specific as you can.

1. What prior experiences have you had that are relevant to your topic or setting? What assumptions about your topic or setting have resulted from these experiences? What goals have emerged from these, or have otherwise become important for your research? How have these experiences, assumptions, and goals shaped your decision to choose this topic, and the way you are approaching this project?
2. What potential advantages do you think the goals, beliefs, and experiences that you described have for your study? What potential disadvantages do you think these may create for you, and how might you deal with these?

EXAMPLE 2.4

Researcher Identity Memo for a Study of Educational Reform in Bolivia

Barbara Noel

There are several layers of personal interest I hold in the topic of educational reform in Bolivia. Probably the most personal is the bilingual/bicultural nature I share with the profile of the Bolivian population. It wasn't until I was well into my adulthood that I recognized how deeply being bilingual has shaped my life consciously and unconsciously. Having spent my childhood in Peru and Mexico, with my bicultural parents (Peruvian mother, very Californian father), I was exposed to Spanish yet grew up speaking English at home and at school. When my family moved to Texas I was 11 and shortly after felt the powerful, sneering attitude to everything Latin American. I and the rest of the family quickly, individually, and without any discussion or conscious inner dialogue spent the next few years carving out the Latino in us and successfully assimilating to the mainstream U.S. culture. I continue to observe this inner battle within my siblings and mother. Fourteen years later, I started speaking Spanish again once I realized the futility and extent of destruction from trying to stamp out one culture in favor of another. Since then, I have turned away from a sort of cultural schizophrenia and have begun to identify where I can integrate the two cultures, consciously choosing what I see as the best of both.

In the Bolivian society, I see the same struggle I personally experienced magnified on a very large scale. I see how for most of the nation's history, one dominant culture has sought to eliminate all the others. It is no accident that forced schooling in an incomprehensible language has produced a population where more than half of the adults over 15 years of age are illiterate. The minds of the indigenous people have also been colonized. They passionately fight for their children to speak only Spanish because this, as they see it, is the only vehicle for attaining political voice and economic security. Many of them desperately seek to assimilate and cut out any traces of "cholo" or Indian in them. Even if they or their children understand an indigenous language, they will act as though they don't understand.

I mostly feel angry as I write about these issues. In a way it is this anger and the subsequent passion for justice that drove me to the field of intercultural, bilingual education. Now I find myself inside a whole country wrestling with the same problems my family and I wrestle with. I must be careful to not project my own journey onto my perception of Bolivian society. I need to seek external validation for my perceptions and ongoing theories about this struggle in Bolivia to avoid painting an inaccurate picture. The confusion for me will come from assuming that my inner lens is the same as the lenses of those with whom I speak.

Writing this memo, I have come to see how my personal base could provide a unique contribution to studying this bilingual/bicultural struggle in Bolivia. My own experience will help me capture my interviewees' stories more vividly and sensitively. By having an inside perspective, I can help the people I interview trust me. I need to figure out just how much to share with them in order to open up dialogue and yet not have my experience corrupt their story. This sort of sharing, "I've been there too," may help my interviewees move past the barrier of how I look, a blond "gringa" from an imperialistic nation.

Another layer of interest in this study is the experience of teachers as they undergo making changes the reform asks them to make. They are being told to completely change their mental schemas for teaching, from a transmission approach to a constructivist approach, without any clear guidelines, models, or examples. This leaves the teachers at a loss as to how to begin. Six years after the reform program began they are still confused. I also entered the profession under similar circumstances, when in the U.S. teachers were being told to teach through a whole language approach. It was like being in a dark room not knowing what to grab on to and trying to act as if you have everything under control lest your job be in jeopardy. Had someone interviewed me about this process at that time, my major concern would have been to appear as if everything is wonderful and that the approach was a magic bullet for teaching. I would have been alienated from my colleagues if they had found out I had said anything remotely negative. This experience helps me to understand how vulnerable these teachers might feel and their need for expressing bravado at all costs.

The personal strength I have in this area is also my biggest weakness. My ability to "put myself in their shoes" and view things from behind their lenses can also get confused by my own projection of the situation based on my own experiences. I might also be tempted to move beyond my role as investigator to reformer, provider of "magic bullets." In the past, I have impulsively offered several workshops, at no cost, just because I'd gotten so caught up in the deep needs I've perceived in their practice and their desire to learn. I need to measure my energies so that I can indeed finish what I start out to do. It will be hard to balance this relationship. I don't feel comfortable just going in as an investigator, yet my "save the world" inspirations need to be tempered into a practical approach that meets the dual purposes of helping and investigating. For me, the reform provides hope that a society may start turning around a long history of oppression by valuing its deeply multicultural character in a way I was able to do on a minute scale.

Addendum, July 2000

It is now several months since I wrote this memo. After having read through it again, I notice several things I learned as a result of going through this exercise. Before writing this memo, I knew I felt intensely drawn to the subject but didn't know why. I felt passionate about righting the wrongs but didn't understand where the motivations were coming from or even that they had a personal basis. Had I not identified my motivations for doing research in this area, I would not have realized how strongly my personal experiences could impact my study. I now realize that even though I try to be very aware, my perceptions will be inevitably colored by my personal background.

It would be easy to fault myself as a researcher by thinking that such an emotional attachment would automatically render me unqualified for such a venture. Yet, through the exercise, I was able to turn the coin around and see the strengths that I also bring through a more empathic stance. While my empathy might help me perceive subtle and important motivations for my informants' responses and behavior, it might also introduce dynamics I unconsciously bring into the situation. I also identified a pattern of behavior I engage in which is to get overly involved with a project so that my emotional connection takes over. I lose my focus and change my role from the one I had objectively started out with. Having identified this pattern, I can, in a way, construct an overhead camera to monitor my actions that might often blink a bright red light to indicate overheating.

What I have come away with from this exercise is clarity of purpose. The real reasons for doing the study. I identified how strongly I felt about the importance of the study personally and professionally. This passion has the possibility, then, to become the engine that sparks my flagging energies and guides me through the blind curves and boring straight stretches of mundane routines during the process of data gathering, transcription, and analysis. I am aware of ways I might possibly corrupt the quality of the information. I also understand how my emotional attachment to the study can be beneficial. This type of reflection helps put in motion a mental machinery that can help monitor my reactions and warn me when I veer off course. Now I see how this memo grounds the rest of the study because it clarifies, energizes, and audits the unique role each researcher brings into the arena.

NOTE

1. In this edition, I have called these "goals" rather than "purposes" in order to more clearly distinguish them from the usual meaning of "purpose" in research methods texts. There, "purpose" refers to the specific objective of a study, for example, "The purpose of this study is to investigate (understand, explore) _______" (Creswell, 1994, p. 59). I see this meaning of "purpose" as more closely connected to the research questions of a study, although distinct from these (Locke, Spirduso, & Silverman, 2000, pp. 45–46).

3

Conceptual Framework

What Do You Think Is Going On?

Biologist Bernd Heinrich (1984, pp. 141–151) and his associates once spent a summer conducting detailed, systematic research on ant lions, small insects that trap ants in pits they have dug. Returning to the university in the fall, Heinrich was surprised to discover that his results were quite different from those published by other researchers. Redoing his experiments the following summer to try to understand these discrepancies, Heinrich found that he and his fellow researchers had been led astray by an unexamined assumption they had made about the ant lions' time frame: Their observations hadn't been long enough to detect some key aspects of these insects' behavior. As he concluded, "even carefully collected results can be misleading if the underlying context of assumptions is wrong" (1984, p. 151).

For this reason, the conceptual framework of your study—the system of concepts, assumptions, expectations, beliefs, and theories that supports and informs your research—is a key part of your design (Miles & Huberman, 1994; Robson, 2002). Miles and Huberman (1994) defined a conceptual framework as a visual or written product, one that "explains, either graphically or in narrative form, the main things to be studied—the key factors, concepts, or variables—and the presumed relationships among them" (p. 18). Here, I use the term in a broader sense that includes the actual ideas and beliefs that you hold about the phenomena studied, whether these are written down or not. This may also be called the "theoretical framework" or "idea context" for the study.

The most important thing to understand about your conceptual framework is that it is primarily a conception or model of what is out there that you plan to study, and of what is going on with these things and why—a tentative *theory* of the phenomena that you are investigating. The function of this theory is to inform the rest of your design—to help you to assess and refine your goals, develop realistic and relevant research questions, select appropriate methods, and identify potential validity threats to your conclusions. It also helps you *justify* your research, something I discuss in more detail in Chapter 7. In this chapter, I discuss the different sources for this theory, and how to use theory effectively in your design. I describe the *nature* of theory in more detail later in the chapter, in dealing with the uses of existing theory. Here, I want to emphasize that your conceptual framework *is* a theory, however tentative or incomplete it may be.

What is often called the "research problem" is a part of your conceptual framework, and formulating the research problem is often seen as a key task in designing your study. It is part of your conceptual framework (although it is often treated as a separate component of a research design) because it identifies something that is *going on* in the world, something that is itself problematic or that has consequences that are problematic. Your research problem functions (in combination with your goals) to *justify* your study, to show people why your research is important. In addition, this problem is something that is not fully understood, or that we don't adequately know how to deal with, and therefore we want more information about it. Not every study will have an explicit statement of a research problem, but every worthwhile research design contains an implicit or explicit identification of some issue or problem, intellectual or practical, about which more information is needed. (The justification of "needed" is where your goals come into play.)

Many writers label the part of a research design, proposal, or published paper that deals with the conceptual framework of a study the "literature review." This can be a dangerously misleading term. In developing your conceptual framework, you should *not* simply summarize some body of theoretical or empirical publications, for three reasons:

1. It can lead to a narrow focus on "the literature," ignoring other conceptual resources that may be of equal or greater importance for your study. As Locke, Spirduso, and Silverman (1993) pointed out, "in any active area of inquiry the current knowledge base is not in the library—it is in the invisible college of informal associations among research workers" (p. 48). This knowledge can be found in unpublished papers, dissertations in progress, and grant applications, as well as in the heads of researchers working in this field. Locke et al. (1993) stated that "the best introduction to the current status of a research area is close association with advisors who know the territory" (p. 49). In addition, an exclusive orientation toward "the literature" leads you to ignore your own experience, your speculative thinking (discussed below in the section titled "Thought Experiments"), and any pilot and exploratory research that you've done.

2. It tends to generate a strategy of "covering the field" rather than focusing specifically on those studies and theories that are particularly *relevant* to your research. Literature reviews that lose sight of this need for relevance often degenerate into a series of "book reports" on the literature, with no clear connecting thread or argument. The relevant studies may be only a small part of the research in a defined field, and may range across a number of different approaches and disciplines.[1] In fact, the most productive conceptual frameworks are often those that integrate different approaches, lines of investigation, or theories that no one had previously connected. Bernd Heinrich used Adam Smith's *The Wealth of Nations* in developing a theory of bumblebee foraging and energy balance that emphasized individual initiative, competition, and a spontaneous division of labor, rather than genetic determination or centralized control (Heinrich, 1979, pp. 144–146, 1984, p. 79).

3. It can lead you to think that your task is simply descriptive—to report what previous researchers have found or what theories have been proposed. In constructing a conceptual framework, your purpose is not only descriptive, but also critical; you need to understand (and clearly communicate in your proposal) what *problems* (including ethical problems) there have been with previous research and theory, what contradictions or holes you have found in existing views, and how your study can make an original contribution to our understanding. You need to treat "the literature" not as an *authority* to be deferred to, but as a useful but fallible source of *ideas* about what's going on, and to attempt to see alternative ways of framing the issues. For good examples of this attitude, see Example 3.2 and the "Context" section of Martha Regan-Smith's proposal (see the Appendix).

Another way of putting this is that the conceptual framework for your research study is something that is *constructed*, not found. It incorporates pieces that are borrowed from elsewhere, but the structure, the overall coherence, is something that *you* build, not something that exists ready-made. It is important for you to pay attention to the existing theories and research that are relevant to what you plan to study, because these are often key sources for understanding what is going on with these phenomena. However, these theories and results are often partial, misleading, or simply wrong. Bernd Heinrich found that many of the ideas about ant lions in the literature were incorrect, and his subsequent research led to a much more comprehensive and well-supported theory of their behavior. You will need to critically examine each idea or research finding to see if it is a valid and useful module for constructing a theory that will adequately inform your study.

This idea that existing theory and research provide "modules" that you can use in your own research was developed at length by Becker (1986, pp. 141–146). As he stated,

> I am always collecting such prefabricated parts for use in future arguments. Much of my reading is governed by a search for such useful modules. Sometimes I know I need a particular theoretical part and even have a good idea of where to find it (often thanks to my graduate training in theory, to say a good word for what I so often feel like maligning). (1986, p. 144)

Before describing the sources of these modules, I want to discuss a particularly important part of your conceptual framework—the research paradigm(s) within which you situate your work.

CONNECTING WITH A RESEARCH PARADIGM

One of the critical decisions that you will need to make in designing your study is the paradigm (or paradigms) within which you will situate your work. This use of the term "paradigm," which derives from the work of the historian of science Thomas Kuhn, refers to a set of very general philosophical assumptions about the nature of the world (ontology) and how we can understand it (epistemology), assumptions that tend to be shared by researchers working in a specific field or tradition. Paradigms also typically include specific methodological strategies linked to these assumptions, and identify particular studies that are seen as exemplifying these assumptions and methods. At the most abstract and general level, examples of such paradigms are philosophical positions such as positivism, constructivism, realism, and pragmatism, each embodying very different ideas about reality and how we can gain knowledge of it. At a somewhat more specific level, paradigms that are relevant to qualitative research include interpretivism, critical theory, feminism, postmodernism, and phenomenology, and there are even more specific traditions within these.

It is well beyond the scope of this book to describe these paradigms and how they can inform a qualitative study; good discussions of these issues can be found in Creswell (1998) and Schram (2003). However, I want to make several points that are relevant to using paradigms in your research design:

1. Although some people refer to "the qualitative paradigm," there are many different paradigms within qualitative research, some of which differ radically in their assumptions and implications (cf. Denzin & Lincoln, 2000; Pitman & Maxwell, 1992). It will be important to your research design (and your proposal) to make explicit which paradigm(s) your work will draw on, since a clear paradigmatic stance helps to guide your design decisions and to justify these decisions. Using an established paradigm allows you to build on a coherent and well-developed approach to research, rather than having to construct all of this yourself.

2. You don't have to adopt in total a single paradigm or tradition. It is possible to combine aspects of different paradigms and traditions, although if you do this, you will need to carefully assess the compatibility of the modules that you borrow from each. Schram (2003, p. 79) gave a valuable account of how he combined the ethnographic and life history traditions in his dissertation research on an experienced teacher's adjustment to a new school and community.

3. Your selection of a paradigm (or paradigms) is not entirely a matter of free choice. You have already made many assumptions about the world, your topic, and how we can understand these, even if you have never consciously examined these. Choosing a paradigm or tradition primarily involves assessing which paradigms best fit with your own assumptions and methodological preferences; Becker (1986, pp. 16–17) made the same point about using theory in general. Trying to work within a paradigm (or theory) that doesn't fit your assumptions is like trying to do a physically demanding job in clothes that don't fit—at best you'll be uncomfortable, at worst it will keep you from doing the job well. Such a lack of fit may not be obvious at the outset; it may emerge only as you develop your conceptual framework, research questions, and methods, since these should also be compatible with your paradigmatic stance. Writing memos is a valuable way of revealing and exploring these assumptions and incompatibilities (cf. Becker, 1986, pp. 17–18).

There are four main sources for the modules that you can use to construct the conceptual framework for your study: (1) your own experiential knowledge, (2) existing theory and research, (3) your pilot and exploratory research, and (4) thought experiments. I will begin with experiential knowledge, because it is both one of the most important conceptual resources and the one that is most seriously neglected in works on research design. I will then deal with the use of existing theory and research in research design, in the process introducing a tool, known as "concept mapping," that can be valuable in developing a conceptual framework for your study. Finally, I will discuss the uses of your own pilot research and "thought experiments" in generating preliminary or tentative theories about your subject.

EXPERIENTIAL KNOWLEDGE

Traditionally, what you bring to the research from your own background and identity has been treated as "bias," something whose influence needs to be *eliminated* from the design, rather than a valuable component of it. This has been true to some extent even in qualitative research, despite the fact that qualitative researchers have long recognized that in this field, the researcher *is* the instrument of the research. In opposition to the traditional view, C. Wright Mills, in a classic essay, argued that

> the most admirable scholars within the scholarly community . . . do not split their work from their lives. They seem to take both too seriously to allow such dissociation, and they want to use each for the enrichment of the other. (1959, p. 195)

Separating your research from other aspects of your life cuts you off from a major source of insights, hypotheses, and validity checks. Alan Peshkin, discussing the role of subjectivity in the research he had done, concluded that

> the subjectivity that originally I had taken as an affliction, something to bear because it could not be foregone, could, to the contrary, be taken as "virtuous." My subjectivity is *the* basis for the story that I am able to tell. It is a strength on which I build. It makes me who I am as a person *and* as a researcher, equipping me with the perspectives and insights that shape all that I do as a researcher, from the selection of topic clear through to the emphases I make in my writing. Seen as virtuous, subjectivity is something to capitalize on rather than to exorcise. (Glesne & Peshkin, 1992, p. 104)

Anselm Strauss emphasized many of the same points in discussing what he called "experiential data"—the researcher's technical knowledge, research background, and personal experiences. He argued that

> These experiential data should not be ignored because of the usual canons governing research (which regard personal experience and data as likely to bias the research), for these canons lead to the squashing of valuable experiential data. We say, rather, "mine your experience, there is potential gold there!" (1987, p. 11)

Students' proposals sometimes seem to systematically ignore what their authors know from their own experience about the settings or issues they propose to study; this can seriously damage the proposal's credibility.

Both Peshkin and Strauss emphasized that this is not a license to uncritically impose one's assumptions and values on the research. Reason (1988, 1994) used the term "critical subjectivity" to refer to

> a quality of awareness in which we do not suppress our primary experience; nor do we allow ourselves to be swept away and overwhelmed by it; rather we raise it to consciousness and use it as part of the inquiry process. (1988, p. 12)

The explicit incorporation of your identity and experience in your research has gained wide theoretical and philosophical support (e.g., Berg & Smith, 1988; Denzin & Lincoln, 2000; Jansen & Peshkin, 1992). The philosopher Hilary Putnam (1987, 1990) argued that there cannot, even in principle, be such a thing as a "God's eye view," a view that is the one true "objective" account. *Any* view is a view *from some perspective,* and therefore is shaped by the location (social and theoretical) and "lens" of the observer.

Philosophical argument does not, however, solve the problem of how to incorporate this experience most productively in your research design, or how to assess its effect on your research. Peshkin's account of how he became aware of the different "I's" that influenced and informed his studies was discussed in Chapter 2, and Jansen and Peshkin (1992) and Grady and Wallston (1988, pp. 40–43) provided valuable examples of researchers using their own subjectivity and experience in their research. At present, however, there are few well-developed and explicit strategies for doing this.

The technique that I call a "researcher identity memo," which was introduced in Chapter 2 for reflecting on your own goals and their relevance for your research, can also be used to explore your assumptions and experiential knowledge. I originally got the idea for this sort of memo from a talk by Robert Bogdan, who described how, before beginning a study of a neonatal intensive care unit of a hospital, he tried to write out all of the expectations, beliefs, and assumptions that he had about hospitals in general and neonatal care in particular, as a way of identifying and taking account of the perspective that he brought to the study. This exercise can be valuable at any point in a study, not just at the outset. Example 3.1 is part of one of my own identity memos, written while I was working on a paper on diversity, solidarity, and community (Maxwell, n.d.), trying to develop a theory that incorporated contact and interaction, as well as shared characteristics, as a basis for community. Example 3.2 is a memo in which the researcher used her own experience to refocus a study of women's use of breast self-examination. Example 2.4, in the previous chapter, deals in part with the author's prior experiences and how these influenced her understanding of educational reform in Bolivia, as well as her goals.

EXAMPLE 3.1

Identity Memo on Diversity

I can't recall when I first became interested in diversity; it's been a major concern for at least the last 20 years. . . . I do remember the moment that I consciously realized that my mission in life was "to make the world safe for diversity"; I was in Regenstein Library at the University of Chicago one night in the mid-'70s talking to another student about why we had gone into anthropology, and the phrase suddenly popped into my head.

However, I never gave much thought to tracing this position any further back. I remember, as an undergraduate, attending a talk on some political topic, and being struck by two students' bringing up issues of the rights of particular groups to retain their cultural heritages; it was an issue that had never consciously occurred to me. And I'm sure that my misspent youth reading science fiction rather than studying had a powerful influence on my sense of the importance of tolerance and understanding of diversity; I wrote my essay for my application to college on tolerance in high school society. But I didn't think much about where all this came from.

It was talking to the philosopher Amelie Rorty in the summer of 1991 that really triggered my awareness of these roots. She had given a talk on the concept of moral diversity in Plato, and I gave her a copy of my draft paper on diversity and solidarity. We met for lunch several weeks later to discuss these issues, and at one point she asked me how my concern with diversity connected with my background and experiences. I was surprised by the question, and found I really couldn't answer it. She, on the other hand, had thought about this a lot, and talked about her parents emigrating from Belgium to the US, deciding they were going to be farmers like "real Americans," and with no background in farming, buying land in rural West Virginia and learning how to survive and fit into a community composed of people very different from themselves.

This made me start thinking, and I realized that as far back as I can remember I've felt different from other people, and had a lot of difficulties as a result of this difference and my inability to "fit in" with peers, relatives, or other people generally. This was all compounded by my own shyness and tendency to isolate myself, and by the frequent moves that my family made while I was growing up. . . .

The way in which this connects with my work on diversity is that my main strategy for dealing with my difference from others, as far back as I can remember, was *not* to try to be more *like* them (similarity-based), but to try to be *helpful* to them (contiguity-based). This is a bit oversimplified, because I also saw myself as somewhat of a "social chameleon," adapting to whatever situation I was in, but this adaptation was much more an *interactional* adaptation than one of becoming fundamentally similar to other people.

It now seems incomprehensible to me that I never saw the connections between this background and my academic work. . . .

[The remainder of the memo discusses the specific connections between my experience and the theory of diversity and community that I had been developing, which sees both similarity (shared characteristics) and contiguity (interaction) as possible sources of solidarity and community (Maxwell, n.d.).]

EXAMPLE 3.2

How One Researcher Used Her Personal Experience to Refocus Her Research Problem

I had spent countless hours in the library, reading the literature on women's practice of breast self-examination (BSE). The articles consisted of some research studies, some editorials in major medical journals, and some essays. The research base was very weak, mainly surveys asking some group of women whether they did BSE, and if not, why not. The groups often were not large or representative. The questions and format varied tremendously from study to study. That most women did not do it was clear, having been found repeatedly. Why they did not do it was not at all clear. I was developing a long list of possible reasons women did not do it. They seemed to fall into three categories: (1) Women were ignorant of how or why to do BSE; (2) women were too modest to touch themselves; and (3) women were too fearful of what they would find. The reasons all seemed quite plausible, but somehow were not satisfactory. The question kept repeating itself, "Why *don't* women do BSE?" Then I asked the question of myself, "Why don't I do BSE?" I knew none of the reasons explained my behavior. Then I changed the question: "What would get me to do it?" It occurred to me that, if a friend called each month and asked if I had done it, I would do it, either in anticipation of her call or immediately afterward. Changing the question to a positive one completely changed my way of thinking about the problem: "What would *encourage* women to do BSE?" The new question opened a range of possibilities by putting BSE in the context of behavior modification, which offered a variety of testable techniques for changing behavior. (Grady & Wallston, 1988, p. 41)

PRIOR THEORY AND RESEARCH

The second major source of modules for your conceptual framework is prior theory and research—not simply published work, but other people's theories and empirical research as a whole. I will begin with theory, because it is for most people the more problematic and confusing of the two, and then deal with using prior research for other purposes than as a source of theory.

I'm using the term "theory" to refer to something that is considerably broader than its usual meaning in discussions of research methods. By "theory," I mean simply a set of concepts and the proposed relationships among these, a structure that is intended to represent or model something about the world. As LeCompte and Preissle (1993, p. 239) stated, "theorizing is simply the cognitive process of discovering or manipulating abstract categories and the relationships among these categories." My only modification of this is to include not simply abstract categories, but concrete and specific concepts as well.

This use encompasses everything from so-called "grand theory," such as behaviorism, psychoanalysis, or rational choice theory, to specific, everyday explanations of a particular event or state, such as "Dora (my 8-year-old daughter) doesn't want to go to school today because she's angry at her teacher for correcting her yesterday." That is, I'm not using "theory" to denote a particular *level* of complexity, abstraction, or generality of explanatory propositions, but to refer to the *entire range* of such propositions. All such explanations have fundamental features in common, and for my purposes the similarities are more important than the differences.[2]

Thus, theory is not an arcane and mysterious entity that at some point in your training you learn to understand and master. As Groucho Marx used to say on the 1950s TV game show *You Bet Your Life,* "It's an ordinary household word, something you use every day." The simplest form of theory consists of two concepts joined by a proposed relationship. Such a theory can be as general as "Positive reinforcement leads to continuation of the reinforced behavior," or as specific as "An asteroid impact caused the extinction of the dinosaurs." The important point is what *makes* this a theory: the linking of two concepts by a proposed relationship.

A major function of theory is to provide a model or map of *why* the world is the way it is (Strauss, 1995). It is a simplification of the world, but a simplification aimed at clarifying and explaining some aspect of how it works. Theory is a statement about what is going on with the phenomena that you want to understand. It is not simply a "framework," although it can provide that, but a *story* about what you think is happening and why. A useful theory is one that tells an enlightening story about some phenomenon, one that gives you new insights and broadens your understanding of that phenomenon. (See the discussion of causal processes in Chapter 2.)

Glaser and Strauss's term "grounded theory" (1967), which has had an important influence on qualitative research, does not refer to any particular *level* of theory, but to theory that is inductively developed during a study (or series of studies) and in constant interaction with the data from that study.

This theory is "grounded" in the actual data collected, in contrast to theory that is developed conceptually and then simply tested against empirical data. In qualitative research, both existing theory and grounded theory are legitimate and valuable.

The Uses of Existing Theory

Using existing theory in qualitative research has both advantages and risks, as discussed earlier. The advantages of existing theory can be illustrated by two metaphors:

Theory is a coat closet. (I got this metaphor from Jane Margolis, who once described Marxism as a coat closet: "You can hang anything in it.") A useful high-level theory gives you a framework for making sense of what you see. Particular pieces of data, which otherwise might seem unconnected or irrelevant to one another or to your research questions, can be related by fitting them into the theory. The concepts of the existing theory are the "coat hooks" in the closet; they provide places to "hang" data, showing their relationship to other data. However, no theory will accommodate all data equally well; a theory that neatly organizes some data will leave other data disheveled and lying on the floor, with no place to put them.

Theory is a spotlight. A useful theory *illuminates* what you see. It draws your attention to particular events or phenomena, and sheds light on relationships that might otherwise go unnoticed or misunderstood. Bernd Heinrich, discussing an incident in his investigation of the feeding habits of caterpillars, stated that

> The clipped leaf stood out as if flagged in red, because it didn't fit my expectations or theories about how I thought things ought to be. My immediate feeling was one of wonder. But the wonder was actually a composite of different theories that crowded my mind and vied with each other for validation or rejection. . . . Had I no theories at all, the partially eaten leaf on the ground would not have been noticed. (1984, pp. 133–134)

By the same token, however, a theory that brightly illuminates one area will leave other areas in darkness; no theory can illuminate everything.

A study that makes excellent use of existing theory is described in Example 3.3.

EXAMPLE 3.3

Using Existing Theory

Eliot Freidson's book *Doctoring Together: A Study of Professional Social Control* (1975) is an account of his research in a medical group practice, trying to understand how the physicians and administrators he studied identified and dealt with violations of professional norms. In conceptualizing what was going on in this practice, he used three broad theories of the social organization and control of work. He referred to these as the entrepreneurial, or physician-merchant, model, deriving from the work of Adam Smith; the bureaucratic, or physician-official, model, deriving to a substantial extent from Max Weber; and the professional, or physician-craftsman, model, which has been less clearly conceptualized and identified than the others. He showed how all three theories provide insight into the day-to-day work of the group he studied, and he drew far-ranging implications for public policy from his results.

Freidson also used existing theory in a more focused (and unexpected) way to illuminate the results of his research. He argued that the social norms held by the physicians he studied allowed considerable differences of opinion about both the technical standards of work performance and the best way to deal with patients. These norms "limited the critical evaluation of colleagues' work and discouraged the expression of criticism" (p. 241). However, the norms also strongly opposed any outside control of the physicians' practice, defining physicians as the only ones capable of judging medical work. "The professional was treated as an individual free to follow his own judgment without constraint, so long as his behavior was short of blatant or gross deficiencies in performance and inconvenience to colleagues" (p. 241). Freidson continued:

> This is a very special kind of community that, structurally and normatively, parallels that described by Jesse R. Pitts as the "delinquent community" of French schoolchildren in particular and French collectivities in general during the first half of the twentieth century. . . . Its norms and practice were such as to both draw all members together defensively against the outside world . . . and, internally, to allow each his freedom to act as he willed. (pp. 243–244)

He presented striking similarities between the medical practice he studied and the French peer group structure identified by Pitts. He coined the phrase "professional delinquent community" to refer to professional groups such as the one he described, and used Pitts's theory to illuminate the process by which this sort of community develops and persists.

However, Becker (1986) warned that the existing literature, and the assumptions embedded in it, can deform the way you frame your research, causing you to overlook important ways of conceptualizing your study or key implications of your results. The literature has the advantage of what he calls "ideological hegemony," so that it is difficult to see any phenomenon in ways that are different from those that are prevalent in the literature. Trying to fit your insights into this established framework can deform your argument, weakening its logic and making it harder for you to see what a new way of framing the phenomenon might contribute. He explained how his own research on marijuana use was deformed by existing theory:

> When I began studying marijuana use in 1951, the ideologically dominant question, the only one worth looking at, was "Why do people do a weird thing like that?" and the ideologically preferred way of answering it was to find a psychological trait or social attribute which differentiated people who did from people who didn't . . . [M]y eagerness to show that this literature (dominated by psychologists and criminologists) was wrong led me to ignore what my research was really about. I had blundered onto, and then proceeded to ignore, a much larger and more interesting question: how do people learn to define their own internal experiences? (1986, pp. 147–148)

I had the same experience with my dissertation research on kinship in an Inuit community in northern Canada. At the time that I conducted the research, the literature on kinship in anthropology was dominated by a debate between two theories of the meaning of kinship, one holding that in all societies kinship was fundamentally a matter of biological relationship, the other arguing that biology was only one possible meaning of kinship terms, another being social relatedness. I framed my dissertation (Maxwell, 1986) in terms of these two theories, arguing that my evidence mainly supported the second of these theories, though with significant modifications. It was only years later that I realized that my research could be framed in a more fundamental and interesting way—What is the nature of relationship and solidarity in small, traditional communities? Are these based on, and conceptualized in terms of, similarity (in this case, biological similarity or shared genetic substance) or social interaction? (See Example 3.1.) My research could have been much more productive if I had grasped this theoretical way of framing the study at the outset.

Becker argued that there is no way to be sure when the dominant approach is wrong or misleading or when your alternative is superior. What you can do is to try to identify the ideological components of the established approach, and to see what happens when you abandon these assumptions. He claimed that "a serious scholar ought routinely to inspect competing ways of talking about the same subject matter," and cautioned, "use the literature, don't let it use you" (1986, p. 149). An awareness of alternative sources of concepts and theories about the phenomena you are studying—including sources other than "the literature"—is an important counterweight to the ideological hegemony of existing theory and research.

There are thus two main ways in which qualitative researchers often fail to make good use of existing theory: by not using it enough, and by relying too heavily and uncritically on it. The first fails to explicitly apply any prior analytic abstractions or theoretical framework to the study, thus missing the insights that only existing theory can provide. Every research design needs *some* theory of the phenomena you are studying, even if it is only a common-sense one, to guide the other design decisions you make. The second type of failure has the opposite problem: It *imposes* theory on the study, shoehorning questions, methods, and data into preconceived categories and preventing the researcher from seeing events and relationships that don't fit the theory.

The imposition of dominant theories is a serious ethical problem, not simply a scientific or practical one (Lincoln, 1990); it can marginalize or dismiss the theories of participants in the research, and conceal or minimize oppression or exploitation of the group studied. (In some cases, the dominant theory is itself ethically problematic, as in the case of theories of the problems that disadvantaged groups encounter that unjustifiably "blame the victim.") To be genuinely qualitative research, a study must take account of the theories and perspectives of those studied, rather than relying entirely on established views or the researcher's own perspective.

The tension between these two problems in applying theory (underuse and overuse) is an inescapable part of research, not something that can be "solved" by some technique or insight. A key strategy for dealing with this is embodied in the scientific method, as well as in interpretive approaches such as hermeneutics: Develop or borrow theories and continually *test* them, looking for discrepant data and alternative ways (including the research participants' ways) of making sense of the data. (I discuss this further in Chapter 6, as a central issue in validity.) Bernd Heinrich described searching for crows' nests, in which you look through the trees for a dark spot against the sky, and then try to see a glimmer of light through it (real crows' nests are opaque): "It was like science: first you look for something, and then when you think you have it you do your best to prove yourself wrong" (1984, p. 28).

CONCEPT MAPS

For many students, the development and use of theory is the most daunting part of a qualitative study. At this point, therefore, I want to introduce a tool for developing and clarifying theory, known as "concept mapping." This was originally developed by Joseph Novak (Novak & Gowin, 1984), first as a way to understand how students learned science, and then as a tool for teaching science. A similar strategy is one that Miles and Huberman (1994, pp. 18–22) called a "conceptual framework." Anselm Strauss (1987, p. 170) provided a third variation, which he called an "integrative diagram." These approaches have so much in common that I will present them as a single strategy, ignoring for the moment some important differences in the way they are used.

Figures 3.1 to 3.5 provide a variety of examples of concept maps; further examples can be found in Miles and Huberman (1994) and Strauss (1987, pp. 170–183).

As these figures illustrate, a concept map of a theory is a visual display of that theory—a picture of what the theory says is *going on* with the phenomenon you're studying. These maps do not depict the study itself, nor are they a specific part of either a research design or a proposal. [However, concept maps *can* be used to visually present the design or operation of a study—my model of research design (Figure 1.1) is just such a map.] Rather, concept mapping is a *tool* for developing the conceptual framework for your design. And like a theory, a concept map consists of two things: concepts and the relationships among these. These are usually represented, respectively, as labeled circles or boxes and as arrows or lines connecting these.

There are several reasons for creating concept maps:

1. To pull together, and make visible, what your implicit theory is, or to clarify an existing theory. This can allow you to see the implications of the theory, its limitations, and its relevance for your study.

2. To *develop* theory. Like memos, concept maps are a way of "thinking on paper"; they can help you see unexpected connections, or identify holes or contradictions in your theory and help you to figure out ways to resolve these.

Concept maps usually require considerable reworking in order to get them to the point where they are most helpful to you; don't expect to generate your final map on the first try. One useful way of developing a concept map is on a blackboard, where you can erase unsuccessful attempts or pieces that don't seem to work well, and play with possible arrangements and connections. (The disadvantage of this is that it doesn't automatically create a "paper trail" of your attempts; such a trail can help you to understand how your theory has changed and avoid repeating the same mistakes.) There are also a

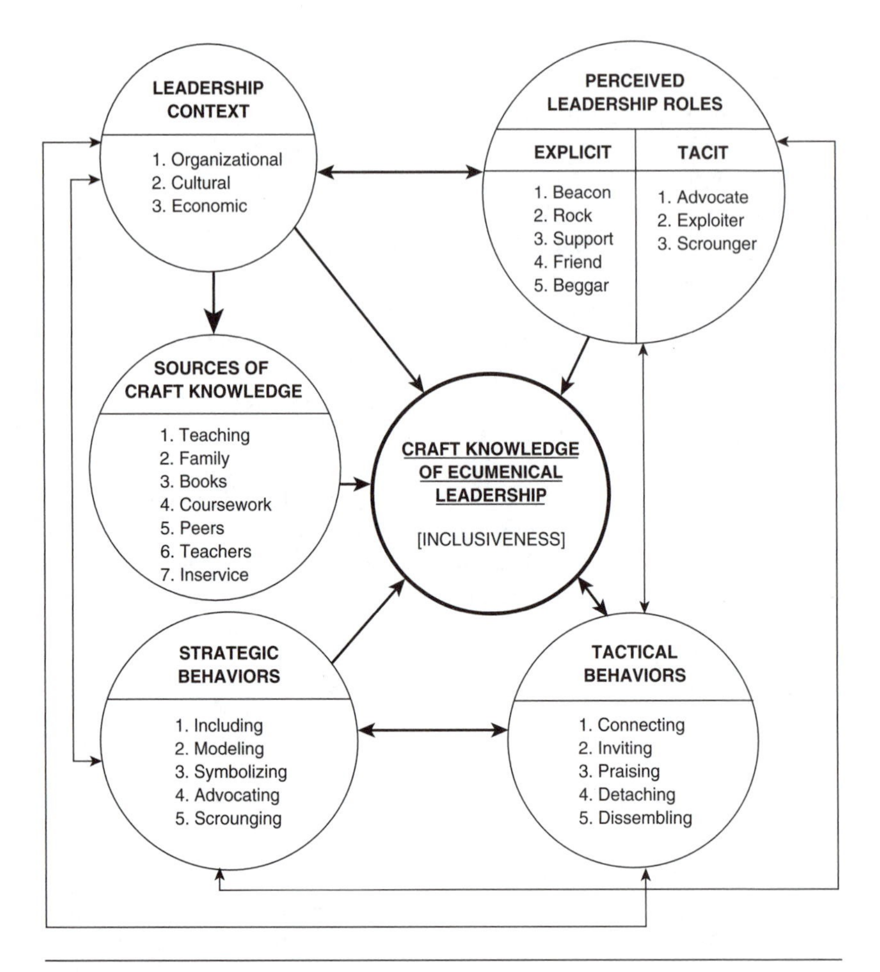

Figure 3.1 A Study of Newfoundland Principals' Craft Knowledge

SOURCE: From "Swamp Leadership: The Wisdom of the Craft," by B. Croskery, 1995, unpublished doctoral dissertation, Harvard Graduate School of Education.

variety of computer programs that can be used to create concept maps (Weitzman & Miles, 1995); I used one of the most popular ones, Inspiration, to create many of the diagrams for this book. Strauss (1987, pp. 171–182) provided a valuable transcript of his consultation with one student, Leigh Star, in helping her to develop a concept map for her research. Exercise 3.1 provides some ways of getting started on creating concept maps of your conceptual framework.

The following factors appear to influence the decisions to keep at home an adult family member who is dependent because of disabilities, rather than "placing" or "institutionalizing" the adult child:

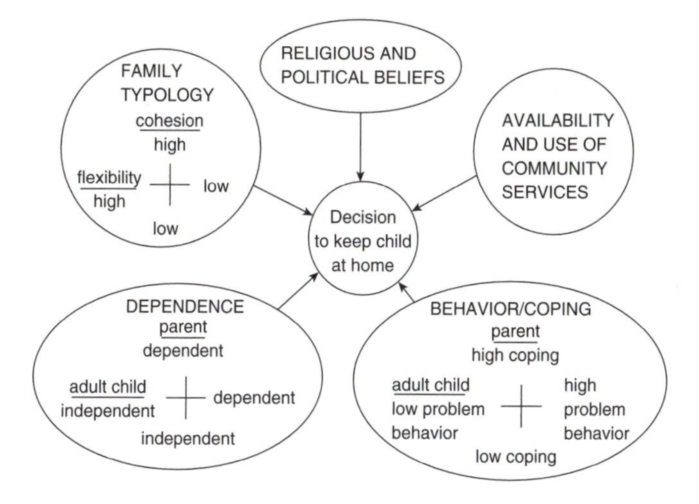

FAMILY TYPOLOGY is a model of intrafamily interactions and permeability of family boundaries developed by David Kantor and expanded by Larry Constantine. Although I have not collected data on family typologies, intuition and existing data favor the prediction that families in the upper-right quadrant (*closed* family systems) and lower-right quadrant (*synchronous* family systems) are more likely to keep the dependent adult child at home, whereas families in the upper-left quadrant (open families) and lower-left quadrant (*random* families) are more likely to place the adult child.

In the DEPENDENCE grid, preliminary data indicate that the upper-left quadrant (high parental dependence, low child dependence) tends to correlate with a decision to keep the adult child at home, whereas the lower-right quadrant (parental independence, high care needs in a child) tends to correlate with placing the adult child.

Similarly, in the BEHAVIOR/COPING grid, the upper-left quadrant (minimal behavior problems, high parental coping) tends to correlate with keeping the adult child at home, whereas the lower-right quadrant (serious behavior problems, low parental coping) tends to correlate with a decision to place the adult child.

Figure 3.2 Factors Affecting the Decision to Keep a Dependent Adult Child at Home

SOURCE: Adapted from "The Families of Dependent Handicapped Adults: A Working Paper," by B. Guilbault, 1989, unpublished manuscript.

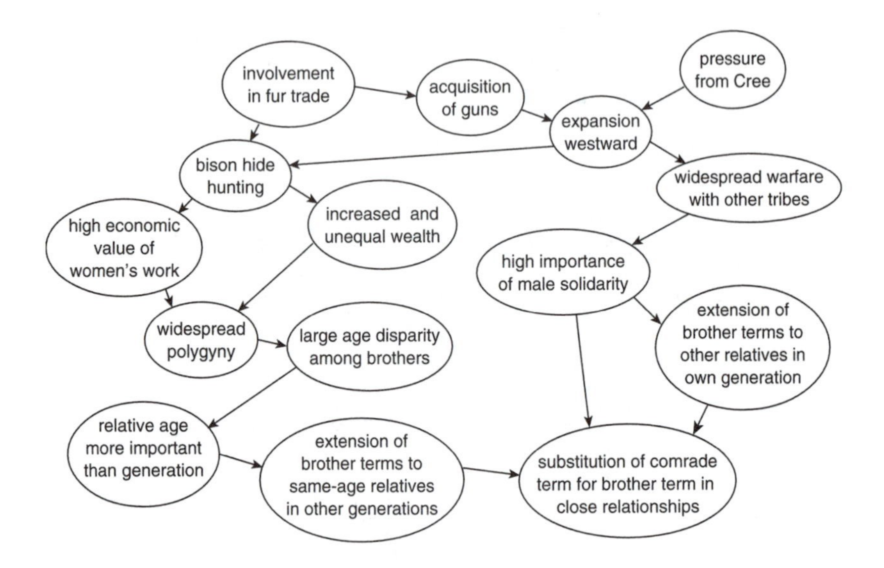

This map displays some of the events and influences leading to the widespread use of "brother" terms in Blackfeet society by the late 1800s. More than any other Plains tribe, the Blackfeet were involved in the fur trade. This led to increased wealth (including guns), a greater value of women's work in preparing bison hides for trade, a highly unequal distribution of wealth that favored men who had many horses for bison hunting, and a massive increase in polygyny, as wealthy men acquired large numbers of wives to process hides. The acquisition of guns and horses allowed the Blackfeet to move westward into the Plains, driving out the tribes that had previously lived there. The increase in warfare and bison hunting created a greater need for male solidarity and led to the widespread use of brother terms between men of the same generation to enhance this solidarity. However, the increased polygyny led to a wider range of ages within a man's generation and to the extension of brother terms to men of other generations who were of about the speaker's age. This proliferation of the use of brother terms eventually diluted their solidarity value, generating a new term, "comrade," which was often used in close relationships between men.

Figure 3.3 Causes of Changes in Blackfeet Kin Terminology

SOURCE: Adapted from "The Development of Plains Kinship Systems," by J. A. Maxwell, 1971, unpublished master's thesis, University of Chicago, and "The Evolution of Plains Indian Kin Terminologies: A Non-reflectionist Account," by J. A. Maxwell, 1978, *Plains Anthopologist, 23*, 13–29.

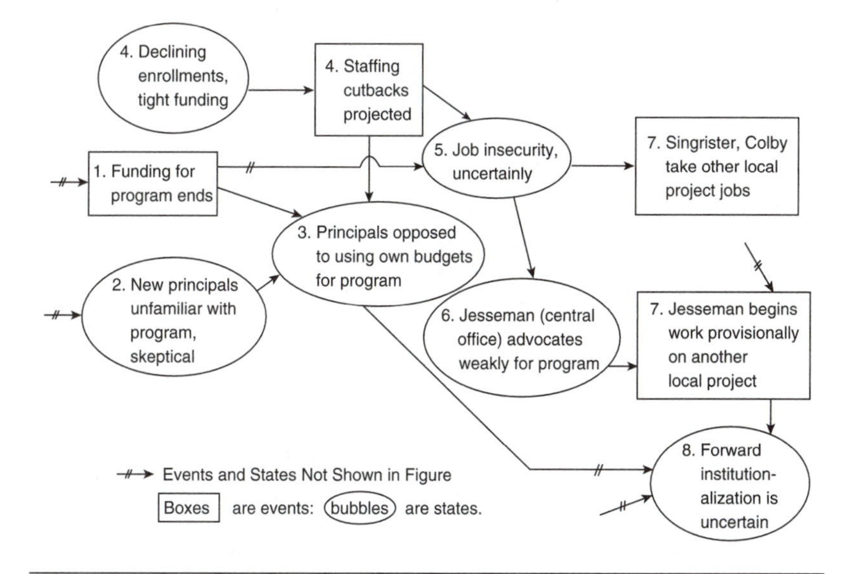

Figure 3.4 Excerpt From an Event-State Network: Perry-Parkdale School

SOURCE: From *Qualitative Data Analysis: An Expanded Sourcebook* (2nd ed.), by M. B. Miles and A. M. Huberman, 1994, Thousand Oaks, CA: Sage.

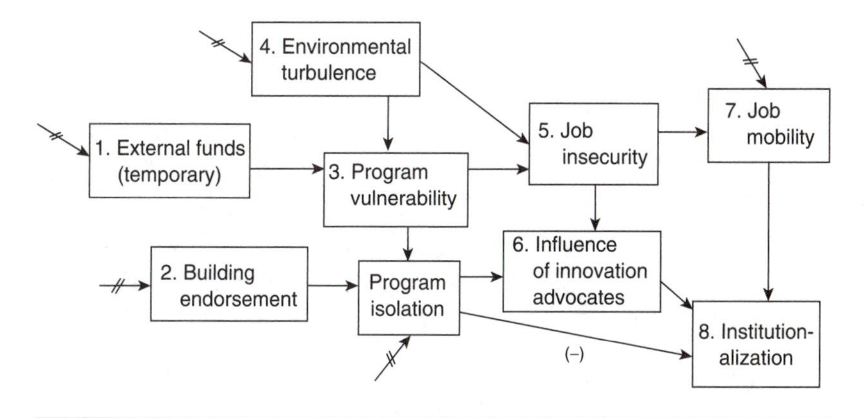

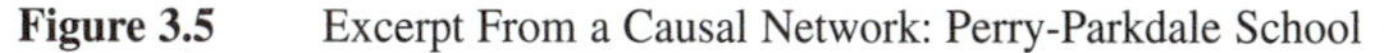

Figure 3.5 Excerpt From a Causal Network: Perry-Parkdale School

SOURCE: From *Qualitative Data Analysis: An Expanded Sourcebook* (2nd ed.), by M. B. Miles and A. M. Huberman, 1994, Thousand Oaks, CA: Sage.

EXERCISE 3.1

Creating a Concept Map for Your Study

How do you develop a concept map? First, you need to have a set of concepts to work with. These can come from existing theory, from your own experience, or from the people you are studying—their *own* concepts of what's going on (discussed below in the section titled "Pilot Research"). The main thing to keep in mind is that at this point you are trying to represent the theory *you already have* about the phenomena you are studying, not primarily to invent a new theory.

If you don't already have a clear conceptual framework for this, there are several strategies you can use to develop your map. Strauss (1987, pp. 182–183) and Miles and Huberman (1994, p. 22) provided additional advice on how to develop concept maps for your study.

1. You can think about the key words you use in talking about this topic; these probably represent important concepts in your theory. You can pull some of these concepts directly from things you've already written about your research.
2. You can take something you've already written and try to map the theory that is implicit (or explicit) in this. (This is often the best approach for people who don't think visually and prefer to work with prose.)
3. You can take one key concept, idea, or term and brainstorm all of the things that might be related to this, then go back and select those that seem most directly relevant to your study.
4. You can ask someone to interview you about your topic, probing for what you think is going on and why; then listen to the tape and write down the main terms you use in talking about it. Don't ignore concepts based on your own experience rather than "the literature"; these can be central to your conceptual framework.

Once you've generated some concepts to work with, ask yourself how these are related. What connections do you see among them? Leigh Star (quoted in Strauss, 1987, p. 179) suggested beginning with one category or concept and drawing "tendrils" to others. What do you think are the important connections between the concepts you're using? The key pieces of a concept map aren't the circles, but the arrows; these represent proposed *relationships* between the concepts or events. Ask yourself the following questions: What do I mean by this particular arrow? What does it *stand for?* Think of *concrete* examples of what you're dealing with, rather than working only with abstractions. Don't lock yourself into the first set of categories you select, or the first arrangement you try. Brainstorm different ways of putting the concepts together; move the categories around to see what works best. Ask questions about the diagram, draw possible connections, and think about whether they make sense.

Finally, write a *narrative* of what this concept map says about the phenomena you are studying. Try to capture in words the ideas that are embodied in the diagram. Figures 3.2 and 3.3 present concept maps with accompanying narratives; Miles and Huberman (1994, pp. 135–136, 159–161) and Strauss (1987, pp. 203–209) provided additional examples. This is an important part of the exercise, and can suggest ways to develop your theory. For example, it can point out when something in your map is simply a placeholder for the actual concept or relationship that you need; Becker (1986) described such placeholders as "meaning nothing in themselves, [but] they mark a place that needs a real idea" (p. 83; he also gave a good example of this on pp. 52–53).

Avoid getting stuck in what Miles and Huberman (1994, p. 22) called a "no-risk" map, in which all the concepts are global and abstract and there are two-directional arrows everywhere. This sort of diagram can be useful as a brainstorming exercise at the beginning, providing you with a conceptual checklist of things that may be important in your research, but at some point, you need to *focus* the theory. It can be useful at some point to narrow your map to two concepts and the relationship between them, as an exercise in focusing on what's most central to your theory. Make *commitments* to what you think is most important and relevant in your theory.

An initial framework often works best with large categories that hold a lot of things that you haven't yet sorted out. However, you should try to differentiate these categories, making explicit your ideas about the relationships among the items in them. One way to start this is by analyzing each one into subcategories, identifying the different *kinds* of things that go into each. (Figure 3.1 does this for the peripheral categories that connect to the core category.) Another way is to *dimensionalize* the categories (Strauss & Corbin, 1990), trying to separate out their different properties. (Figure 3.2 does this for several of the categories.)

How do you know whether something is a category or a relationship? This is not an easy question to answer; I do this rather intuitively. In fact, many things can be seen as either; there is no one right concept map for the phenomena you're studying, and different maps incorporate different understandings of what's going on. You should try out *alternative* maps for the theory you are developing, rather than sticking rigidly with one formulation. There are also different *kinds* of concept maps, with different purposes; these include:

a. an abstract framework mapping the relationship among concepts

b. a "flowchart"-like account of events and how you think these are connected

c. a causal network of variables or influences

d. a treelike diagram of the meanings of words (e.g., Miles & Huberman, 1994, p. 133)

e. a Venn diagram, representing concepts as overlapping circles (e.g., Miles & Huberman, 1994, p. 249)

You can use more than one of these in a given study; the bottom line is their *usefulness* to you in advancing your understanding of what's going on. Most of Miles and Huberman's examples are best suited to studies of social processes; they aren't necessarily the most useful models for a study of meanings and their relationship to one another. Remember that a concept map is not an end in itself; it is a *tool* for developing theory and making that theory more explicit. Also, keep in mind that a concept map is not something that you do once and are finished with; you should go back and rework your concept maps as your understanding of the phenomena you are studying develops. Be careful of making your map too elegant; this may be the visual equivalent of what Becker called "classy writing" (1986, p. 28), and suggests that you may be emphasizing presentation at the expense of insight.

Different authors use concept maps in different ways. Novak and Gowin took a very diffuse approach—their concepts and relationships could be almost anything, and they labeled their connections in order to keep these clear. Miles and Huberman, on the other hand, were much more focused—their connections generally referred to causal relationships or influences. My advice is to aim for something in between. You can start with a fairly diffuse map, but you should work to focus it and to make it a map of a real *theory* of what's going on.

A key distinction, but one that you may not want to think about until *after* you've developed an initial concept map, is the difference between *variance* maps and *process* maps. (See Chapter 2 on the distinction between variance theory and process theory.) One way to tell the difference is that a variance map usually deals with abstract, general concepts that can take different values, and is essentially timeless; it depicts a general causal or correlational relationship between some factors or properties of things, which are conceptualized as variables. A process map, on the other hand, tells a story; there is a beginning and an end, and the concepts are often specific events or situations, rather than variables.[3] Many students create a variance map in their first attempt at concept mapping, because this is their idea of what theory "ought to" look like, even if their research questions are "how" questions that cry out for a process theory. Figures 3.2 and 3.5 are variance maps, while Figures 3.3 and 3.4 are process maps.

OTHER USES OF EXISTING RESEARCH

A review of prior research can serve many other purposes besides providing you with existing theory (cf. Strauss, 1987, pp. 48–56). Locke, Silverman, and Spirduso (1998) gave a clear and detailed explanation of how to read research publications for a variety of useful tools and resources, which they describe as "finding valuables in research reports" (p. 9). These "valuables" include new terminology, including key words to use in searches; references to other publications and researchers; ways of framing research questions, describing the research, or presenting theory, results, or conclusions; and identification of validity issues and ways of dealing with these. Students often overlook such information in their literature reviews, not seeing its value for their research. You need to learn to read for all of these types of information, and to use these in designing your research.

I would emphasize four specific things, in addition to theory, that prior research can contribute to your research design. First, it can help you to develop a *justification* for your study—to show how your work will address an important need or unanswered question. Martha Regan-Smith used prior research on medical school teaching in this way in her proposal (see the Appendix), showing why the topic she planned to study was important, and demonstrating that previous studies had not answered the specific questions she posed. Such a justification connects your plans to your goals for doing the study (Chapter 2), and I discuss this in more detail in Chapter 7, as part of creating an argument for your research proposal.

Second, prior research can inform your decisions about *methods,* suggesting alternative approaches or revealing potential problems and their solutions. Don't skip over the methods sections of papers; see if what the authors did makes sense, if there were problems with their study that bring their results into question, and if you can use any of their strategies or methods for your own study. If you need more information on what they did, contact the authors; they will usually be glad to help you.

Third, prior research can be a source of *data* that can be used to test or modify your theories. You can see if existing theory, pilot research, or your experiential understanding are supported or challenged by previous results.

Doing this will often require thinking through the *implications* of your theory or understanding to see if these are consistent with others' findings. This is one example of a "thought experiment," which I discuss later in this chapter.

Finally, prior research can help you *generate* theory. Bernd Heinrich, while conducting his thesis research on thermoregulation in sphinx moths (1984, pp. 55–68), discovered that his experimental finding that these moths maintain a constant body temperature while flying was directly contradicted by others' research. He described his response as follows:

> As a first step in my decision to proceed, I spent a few months in the library reading about insect physiology in general and everything about sphinx moths in particular. Something in the known physiology and morphology might provide a clue. It would be necessary to collect more and more details on the problem until I could visualize it as closely as if it were a rock sitting in the palm of my hand. I wanted to find out *how* the moths were thermoregulating. . . .
>
> I came across an obscure French paper of 1919 by Franz Brocher on the anatomy of the blood circulatory system in sphinx moths. The odd thing about these moths is that the aorta makes a loop through their thoracic muscles. In many or most other insects, it passes *underneath* these muscles. . . . (Heinrich, 1984, pp. 63–64)

This paper gave Heinrich the critical clue to how these moths were regulating their body temperature: They were shunting blood through the thoracic muscles (which move the moths' wings) to cool these muscles, which would otherwise overheat, and then losing the excess heat from the abdomen, in the same way that a car's water pump and radiator cool the engine. This theory was confirmed by subsequent experiments.

It is possible, of course, to become *too* immersed in the literature; as C. Wright Mills warned, "you may drown in it . . . Perhaps the point is to know when you ought to read, and when you ought not to" (1959, p. 214). One of Mills's main ways of dealing with this problem was, in reading, to always be thinking of empirical studies that could test the ideas he gained from the literature, both as preparation for actual research and as an exercise of the imagination (1959, p. 205). These two strategies connect to the final two sources for your conceptual framework: pilot studies and thought experiments.

PILOT AND EXPLORATORY STUDIES

Pilot studies serve some of the same functions as prior research, but they can be focused more precisely on your own concerns and theories. You can design pilot studies specifically to test your ideas or methods and explore their implications, or to inductively develop *grounded* theory. What Light, Singer, and Willett (1990) said about an illustrative quantitative study is equally true for qualitative research: "Many features of their design could not be determined without prior exploratory research" (p. 212). And they argued that

> No design is ever so complete that it cannot be improved by a prior, small-scale exploratory study. Pilot studies are almost always worth the time and effort. Carry out a pilot study if *any* facet of your design needs clarification. (1990, p. 213)

Example 3.4 describes how Carol Kaffenberger, whose decision to study adolescent cancer survivors and their siblings was presented in Example 2.1, used a pilot study to help design her dissertation research.

EXAMPLE 3.4

How a Student Used a Pilot Study to Help Design Her Dissertation Research

Following her decision to change her dissertation topic, and a review of the literature on her new topic, Carol Kaffenberger decided to conduct a pilot study to help her plan her dissertation research. She chose to use her own family for this pilot study, for several reasons. First, she wanted to practice her interviews, and believed that her family would provide good feedback and suggestions about her methods and what it was like to be a participant in such a study. Second, she wanted to get a better understanding of the meaning of the cancer experience for her own family (one of the personal goals of her research), and to test her own assumptions about this experience. Third, for personal reasons, she wanted her children to have firsthand knowledge of the work she was about to begin. Finally, her family was a convenient choice, and wouldn't require her to find and gain approval from other families.

Carol learned several valuable things from this pilot study. First, she found that she needed to revise her interview guide, adding questions about issues that she hadn't realized were important, such as family relationships before the diagnosis, the support siblings received during diagnosis and treatment, and how they thought the experience would affect their future. She also discovered additional useful questions, such as asking participants to describe specific events that illustrated what they had been saying. Second, she gained a deeper understanding of her children's experiences, modifying her conceptual framework. Both previous research and her prior beliefs had led her to underestimate the long-term consequences of the cancer experience for her family. She learned that she needed to step back and listen to participants' experiences in new ways. Finally, she found that her own children's responses were sometimes guarded and predictable, due to the consequences of what they said for family relationships, and tended to minimize negative feelings or blame. Although the pilot study was valuable, it could not fully answer the questions she had (Kaffenberger, 1999).

One important use that pilot studies have in qualitative research is to develop an understanding of the concepts and theories held by the people you are studying—what is often called "interpretation." This is not simply a source of additional concepts for your own theory, ones that are drawn from the language of participants; this is a type of concept that Strauss (1987, pp. 33–34) called "in-vivo codes." More important, it provides you with an understanding of the *meaning* that these phenomena and events have for the people who are involved in them, and the perspectives that inform their actions. These meanings and perspectives are not theoretical abstractions; they are real, as real as people's behavior, though not as directly visible. People's ideas, meanings, and values are essential parts of the situations and activities you study, and if you don't understand these, your theories about what's going on will often be incomplete or mistaken (Maxwell, 2004a; Menzel, 1978). In a qualitative study, these meanings and perspectives should constitute a key component of your theory; as discussed in Chapter 2, they are one of the things your theory is *about,* not simply a source of theoretical insights and building blocks for the latter. In Example 3.2, the norms and values held by the physicians studied by Freidson were a major part of what was going on in the medical practice, and are fundamental to his theory. Such meanings and perspectives are also key components of all of the previous examples of concept maps (Figures 3.1 to 3.5). Even in Figure 3.5, in which the concepts are mostly stated in terms of behavior or contextual influences, "job insecurity" refers to *perceived* insecurity; if participants were unaware that their jobs might be eliminated, their behavior wouldn't be affected.

THOUGHT EXPERIMENTS

Thought experiments have a long and respected tradition in the physical sciences (much of Einstein's work was based on thought experiments) and are regularly used in social sciences such as economics, but have received little attention as an explicit technique in discussions of research design, particularly qualitative research design. The best guide to thought experiments in the social sciences that I know of is that of Lave and March (1975), who used the phrase "speculative model building" for this concept. Don't be intimidated by the word "model"; models are no more esoteric than theory, and Lave and March defined "model" as "a simplified picture of a part of the real world" (p. 3). They described their book as "a practical guide to speculation," and provided a detailed introduction to the development and use of speculative models of some process that could have produced an observed result. Although the orientation of their later chapters is mainly quantitative, the first three chapters are very readable and extremely useful for qualitative researchers. Lave and March stated,

> We will treat models of human behavior as a form of art, and their development as a kind of studio exercise. Like all art, model building requires a combination of discipline and playfulness. It is an art that is learnable. It has explicit techniques, and practice leads to improvement. (1975, p. 4)

Thought experiments challenge you to come up with plausible explanations for your and others' observations, and to think about how to support or disprove these. They draw on both theory and experience to answer "what if" questions, and to explore the logical implications of your models, assumptions, and expectations of the things you plan to study. They can both generate new theoretical models and insights, and test your current theory for problems; in fact, *all* theory building involves thought experiments to some extent. They encourage creativity and a sense of discovery, and can help you to make explicit the experiential knowledge that you already possess. Example 3.5 is an illustration of this kind of speculative thinking, and Exercise 3.2 (based on one of Lave and March's examples) provides a simple problem on which to practice your speculative skills. According to Lave and March, "the best way to learn about model building is to do it" (1975, p. 10).

EXAMPLE 3.5

Using a Thought Experiment to Develop a Theory of the Persistence of Illiteracy

One of my students, doing research on illiteracy in the Middle East, used the concept of "cycle of illiteracy" in a memo explaining the persistence of illiteracy in parts of this area. This concept has a certain immediate plausibility—illiterate parents are much more likely to have illiterate children than are literate parents. However, my first reaction to the memo was to perform a thought experiment—to try to think of a *process* by which illiteracy in one generation would create illiteracy in the next generation. Lack of reading materials in the home would have some impact, as might parental values regarding literacy. However, none of these seemed powerful enough to reproduce illiteracy at a time when most children have access to schooling. On the other hand, I *could* easily imagine (and support with data that this student had presented) a cycle of *poverty,* in which poor, illiterate families would be under great pressure to keep their children out of school to work in the home or in farming, depriving them of their main opportunity to learn to read and write. As a result, these children's lack of schooling would make it difficult for them to get jobs that would enable them to escape from poverty, thus recreating the conditions that led to their own illiteracy. This theory suggests that reducing poverty would have a major impact on illiteracy. It also implies that research on the causes of illiteracy needs to address the role of economic factors.

EXERCISE 3.2

Creating a Model of the Development of Friendship Patterns

Suppose we were interested in patterns of friendship among college students. Why are some people friends and not others? We might begin by asking all of the residents of single rooms along a particular dormitory corridor to give us a list of their friends. These lists of friends are our initial data, the results we wish to understand.

If we stare at the lists for a while, we eventually notice a pattern in them. Friends tend to live close to one another; they tend to have adjacent dormitory rooms. What process could have produced this pattern of friendship?

STOP AND THINK. Take a minute to think of a possible process that might produce this observed result.

One *possible* process that might have led to this result is that students can choose their dormitory rooms, and that groups of friends tend to choose adjacent rooms. This process is a speculation about the world. *If* the real world were like our model world, the observed facts should match the model's prediction. Thus, we have found a model, a process, that accounts for our results.

We do not stop here, however. We next ask what other implications this model has. For one, it implies that students in each dormitory friendship group must have known one another previously; thus, they must have attended the university the previous year; thus, there will be fewer friendship clusters among freshmen.

A survey of both a freshman dorm and a junior-senior dorm shows that there are as many friendship clusters among freshmen as among juniors and seniors. This would not be predicted by our model, unless the students knew one another in high school. However, examining the backgrounds of the freshmen shows that almost all of them come from different high schools.

So our model does not do a very good job of explaining what we observed. Some process other than mutual selection by prior friends must be involved. So we try to imagine another process that could have led to these results. Our new speculation is that most college students come from similar backgrounds, and thus have enough in common that they could become friends. Pairs of students who live near each other will have more opportunities for interaction, and are more likely to discover these common interests and values, thus becoming friends. This new speculation explains the presence of friendship clusters in freshman dorms as well as in junior-senior dorms.

STOP AND THINK. What other implications does this model have that would allow you to test it? *How* would you test it?

One implication is that since the chance of contact increases over time, the friendship clusters should become larger as the school year progresses. You could test this by surveying students at several different times during the year. If you did so and discovered that the prediction was correct, the model would seem more impressive. (Can you think of other testable implications?)

—Adapted from Lave and March (1975, pp. 10–12).

One issue that Lave and March's example does *not* deal with is the possibility of alternative models that *also* predict most of the same things as the model you have developed. This is one of the most challenging aspects of model building, and the source of a common flaw in theoretical modeling—accepting a model that successfully predicts a substantial number of things, without seriously attempting to come up with alternative models that would make the same (or better) predictions. For example, Lave and March make an assumption, a widespread one in modern Western societies, that friendship is necessarily based on common characteristics—shared interests and values. An alternative model would be one that abandons this assumption, and postulates that friendship can be created by interaction itself, and not necessarily by common characteristics (see Example 3.1).

STOP AND THINK. What tests could distinguish between these two models?

One possible test would be to investigate the beliefs, interests, and values of freshman dormitory students at both the beginning and the end of the year, to see if pairs of friends consistently had more in common at the beginning of the year than did pairs of students in the same dorm who did *not* become friends. (Determining this similarity at the beginning of the year addresses a possible alternative explanation for greater similarity of beliefs and interests within friendship pairs—that this similarity is a *result* of their friendship, rather than a cause.) If you find that pairs of friends did *not* consistently have more in common than pairs of nonfriends, then Lave and March's model seems less plausible (at least without modification), because it predicts that friends will have more in common than nonfriends. My alternative model *does* predict the observed result, and therefore would deserve further consideration and testing. Eventually, you might develop a more complex model that incorporates both processes.

All of the tests described previously (and the standard approach to model testing in general) are based on variance theory—measuring selected variables to see if they fit the model's predictions. However, there is a much more direct way to test the model—*investigate the actual process,* rather than just its predicted consequences (Menzel, 1978, pp. 163–168). For example, you might do participant observation of student interactions at the beginning of the year, looking at how friendships originate, or interview students about how they became friends with other students. This realist, process-oriented approach to model testing is much better suited to qualitative research than is predicting outcomes (Maxwell, 2004a, 2004c).

Experience, prior theory and research, pilot studies, and thought experiments are the four major sources of the conceptual framework for your study. Putting together a conceptual framework from these sources is a unique process for each study, and specific guidelines for how to do this are not of much use; you should look at examples of others' conceptual frameworks to see how they have done this. The main thing to keep in mind is the need for integration of these components with one another, and with your goals and research questions. The connections between your conceptual framework and your research questions will be taken up in the next chapter.

NOTES

1. For a more detailed explanation of this point, see Locke, Spirduso, and Silverman (2000, pp. 68–69). One qualification to this principle is needed for the "literature review" in a dissertation or dissertation proposal. Some advisors or committee members see this as a demonstration that you know the literature in the field of your study, relevant or not. If you are in this situation, your literature review will need to be more comprehensive than I describe. However, you *still* need to identify the work that is most relevant to your study and the specific ideas that you can use in your conceptual framework (and other aspects of your design), because doing this is essential to creating a coherent presentation of, and argument for, your research plans.
2. For a detailed account of the ways in which researchers can use theory in formulating their goals, research questions, and methods, see LeCompte and Preissle (1993, pp. 115–157).
3. Miles and Huberman tended to refer to variance maps as "causal networks," and to process maps as "event-state networks" (1994, pp. 101–171). This incorrectly equates causal analysis with variance analysis; process analysis can *also* be causal, as discussed in Chapter 2 (cf. Maxwell, 2004a).

4

Research Questions

What Do You Want to Understand?

Your research questions—what you specifically want to understand by doing your study—are at the heart of your research design. They are the one component that directly links to all of the other components of the design. More than any other aspect of your design, your research questions will have an influence on, and should be responsive to, every other part of your study.

In many works on research design (e.g., Light, Singer, & Willett, 1990), research questions are presented as the *starting point* and primary determinant of the design. Such approaches don't adequately represent the interactive and inductive nature of qualitative research. Certainly, *if* you already have well-grounded, feasible research questions that are *worth* answering (and this implies goals and knowledge that justify these questions), the rest of your design (especially your methods and conceptual framework) should be selected and constructed to address these questions. In qualitative research, however, you usually can't come up with such questions without making use of the other components of your design. Locking onto your research questions before having a pretty good sense of what your theoretical and methodological commitments and options are, and the implications of these for your questions, creates the danger of what quantitative researchers call a "Type III error"—answering the wrong question.

For this reason, qualitative researchers often don't develop their eventual research questions until they have done a significant amount of data collection and analysis. (See Example 4.1 and Weiss, 1994, pp. 52–53.) This doesn't mean that qualitative researchers begin a study with *no* questions, simply going into their research with an "open mind" and seeing what is there to be investigated. As discussed in the previous two chapters, every researcher begins with certain goals and a substantial base of experience and theoretical knowledge, and these inevitably highlight certain problems or issues and generate questions about these. These early, provisional questions frame the study in important ways, guide decisions about methods, and influence (and are influenced by) the conceptual framework, preliminary results, and potential validity concerns. My point is that well-constructed, focused questions are generally the *result* of an interactive design process, rather than being the starting point for developing a design.

EXAMPLE 4.1

The Development of Research Questions

Suman Bhattacharjea's dissertation (1994) dealt with the ways in which the female administrators in an educational district office in Pakistan defined, implemented, and controlled their professional tasks and working environment in a gender-segregated and male-dominated society. She began her fieldwork with a single broad question: "What do staff in this office do every day, and who does what?" Her position as a consultant to a computer implementation project required her to spend much of her time interacting with the women in this office; the fact that she was female, spoke virtually the same language, and (being from India) was familiar with some aspects of their situation led to acceptance and trust. When she submitted her dissertation proposal, a year after she began the research, she had focused her study on two specific questions:

1. What is the nature of the expectations that affect female administrators' actions?
2. What strategies do female administrators adopt to deal with these constraints in the context of a gender-segregated and male-dominated environment?

On the basis of the research she had already done, she was able to formulate three propositions as tentative answers to these questions:

1. Female administrators' actions reflect their desire to *maintain harmony* between their roles as women in a gender-segregated society and their roles as officials within a bureaucracy.
2. The major strategy female administrators use in this regard is to try to create a "family-like" environment at work, interacting with their colleagues in ways that parallel their interactions in a domestic setting and thus blurring the distinction between "public" and "private."
3. The implications of this strategy for female administrators' actions depend on the *context* of their interaction, in particular, where this context lies on the "public/private" continuum. Women use different strategies when interacting with other women (most "private" or family-like), with male colleagues within the office, and with other men (least "private" or family-like).

In this chapter, I will discuss the *purposes* that research questions can accomplish in a research design, consider the *kinds* of questions that a qualitative study can best investigate, and give some suggestions on how you can *develop* appropriate and productive research questions.

THE FUNCTIONS OF RESEARCH QUESTIONS

In your research *proposal,* the function of your research questions is to explain specifically what your study will attempt to learn or understand. In your research *design,* the research questions serve two other vital functions: to help you to focus the study (the questions' relationship to your goals and conceptual framework) and to give you guidance for how to conduct it (their relationship to methods and validity) (cf. Miles & Huberman, 1994, pp. 22–25).

A design in which the research questions are too general or too diffuse creates difficulties both in conducting the study—in knowing what sites or participants to choose, what data to collect, and how to analyze these data—and in clearly connecting these to your goals and conceptual framework. If your questions remain on the "What's going on here?" level, you have no clear guide in deciding what data to collect, in selecting or generating relevant theory for your study, or in knowing if your study is meeting your goals. More precisely framed research questions, in contrast, can point you to specific areas of theory that you can use as "modules" in developing an understanding of what's going on, and suggest ways to do the study.

On the other hand, it is possible for your questions to be *too* focused; they may create tunnel vision, leaving out things that are important to the goals (both intellectual and practical) of the study. Research questions that are precisely framed too early in the study may lead you to overlook areas of theory or prior experience that are relevant to your understanding of what is going on, or cause you to not pay enough attention to a wide range of data, data that can reveal important and unanticipated phenomena and relationships.

A third potential problem is that you may be smuggling unexamined assumptions into the research questions themselves, imposing a conceptual framework that doesn't fit the reality you're studying. A research question such as "How do teachers deal with the experience of isolation from their colleagues in their classrooms?" assumes that teachers *do* experience such isolation. Such an assumption needs to be carefully examined and justified, and a question of this form may be better placed as a subquestion to broader questions about the nature of classroom teachers' experience of their work and their relations with colleagues.

Fourth, there is the possibility, an unfortunate but not unknown one with students beginning to write dissertation proposals, that the stated research questions bear little discernable relationship to the students' actual goals and beliefs about what's going on. Instead, they are constructed to satisfy what the student thinks research questions *should* look like, or what committee members will want to see in the proposal. Such questions may be inconsistent with other parts of the design. (See the discussion of Potemkin villages in Chapter 7.) In qualitative research, such questions are often the result of adopting quantitative research conventions for framing questions, conventions that are inappropriate for a qualitative study.

For all of these reasons, there is a real danger in not carefully formulating your research questions in connection with the other components of your design. Your research questions need to take account of why you want to do the study (your goals), your connections to a (or several) research paradigm(s), and what is already known about the things you want to study and your tentative theories about these phenomena (your conceptual framework). You don't want to pose questions for which the answer is already available, that don't clearly connect to what you think is actually going on, or that, even if you answer them, won't advance your goals.

Similarly, your research questions need to be answerable by the kind of study you could actually conduct. You don't want to pose questions that no feasible study could answer, either because the data that might answer them could not be obtained or because the conclusions you might draw would be subject to serious validity threats. These issues will be covered in more detail in the next two chapters.

To develop appropriate research questions for your study, you need to understand clearly what a research question is, and the different kinds of research questions that you might construct. I will first discuss the nature of research questions in general, and then introduce some specific distinctions among research questions that are important for qualitative studies.

RESEARCH QUESTIONS AND OTHER KINDS OF QUESTIONS

A common problem in developing research questions is confusion between intellectual issues—what you want to *understand* by doing the study—and practical issues—what you want to *accomplish.* According to LeCompte and Preissle, "distinguishing between the purpose and the research question is the first problem" in coming up with workable research questions (1993, p. 37). As discussed in Chapter 2, practical concerns usually can't be *directly* addressed by a research study; they are generally best kept as part of your goals. These practical goals should *inform* your research questions, but normally shouldn't be directly *incorporated* into these questions. Instead, you should frame your research questions so that they point you to the information and understanding that will help you to *accomplish* your practical goals or develop the practical implications of what you learn.

A second distinction, one that is critical for interview studies, is that between *research* questions and *interview* questions. Your research questions identify the things that you want to understand; your interview questions generate the data that you need to understand these things. These are rarely the same; the distinction is discussed in more detail in Chapter 5.

RESEARCH HYPOTHESES IN QUALITATIVE DESIGNS

Research questions are not the same as research hypotheses. Research questions state what you want to learn. Research hypotheses, in contrast, are a statement of your tentative answers to these questions, what you think is going on; these answers are normally implications of your theory or experience. The use of explicit research hypotheses is often seen as incompatible with qualitative research. My view, in contrast, is that there is no inherent problem with formulating qualitative research hypotheses; the difficulty has been partly a matter of terminology and partly a matter of the inappropriate application of quantitative standards to qualitative research hypotheses.

Many qualitative researchers explicitly state their ideas about what is going on as part of the process of theorizing and data analysis. These may be called "propositions" rather than hypotheses (Miles & Huberman, 1994, p. 75), but they serve the same function. The distinctive characteristic of hypotheses in qualitative research is that they are typically formulated *after* the researcher has begun the study; they are "grounded" (Glaser & Strauss, 1967) in the data and are developed and tested in interaction with them, rather than being prior ideas that are simply tested against the data (see Example 4.1).

This runs counter to the view, widespread in quantitative research, that unless a hypothesis is framed in advance of data collection, it can't be legitimately tested by the data. This requirement is essential for the *statistical* test of a hypothesis; if the hypothesis is framed after seeing the data, the assumptions of the statistical test are violated. Colloquially, this is referred to as a "fishing expedition"—searching through the data to find what seem to be significant relationships. However, qualitative researchers rarely engage in statistical significance testing, so that this argument is largely irrelevant to qualitative research. "Fishing" for possible answers to your questions is perfectly appropriate in qualitative research, as long as these answers are then *tested* against new evidence and possible validity threats.

The main risk in explicitly formulating hypotheses is that, like prior theory, they can act as blinders, preventing you from seeing what's going on. You should regularly re-examine these hypotheses, asking yourself what alternative ways there are of making sense of your data; thought experiments (Chapter 3) are a good way to do this.

I next want to discuss three specific distinctions among kinds of research questions, ones that are important to consider in developing the questions for your study. These distinctions are between generic and particularistic questions, between instrumentalist and realist questions, and between variance and process questions.

GENERIC QUESTIONS AND PARTICULARISTIC QUESTIONS

There is a widespread, but often implicit, assumption, especially in quantitative research, that research questions should be framed in general terms, and then "operationalized" by means of specific sampling and data collection decisions. For example, there is a tendency to state a research question about racial and ethnic differences in a school as "How do students deal with racial and ethnic difference in multiracial schools?" and to then answer this by studying a particular school as a "sample" from this population of schools, rather than to state the question at the outset as "How do students at North High School deal with racial and ethnic difference?" I will refer to these two types of questions as generic and particularistic questions, respectively.

The assumption that questions should be stated in generic terms may derive in part from the views of logical positivism, in which causal explanation was seen as inherently involving general laws, and the goal of science was to discover such laws. However, this assumption has been challenged by realist philosophers who argue for researchers' ability to observe causation in single cases (Maxwell, 2004a). It also does not fit a great deal of research in the social sciences and in fields such as education, where particularistic questions can be appropriate and legitimate. It is especially misleading in applied research, where the focus is usually on understanding and improving some particular program, situation, or practice.

These two types of questions are linked to the difference between a sampling approach and a case study approach to research. In a sample study, the researcher states a generic question about a broad population, and then selects a particular sample from this population in order to answer the question. In a case study, in contrast, the researcher often selects the case and then states the questions in terms of the particular case selected. A sample study justifies the sampling strategy as a way of attaining what Cook and Campbell (1979) called "statistical conclusion validity"—the representativeness of the specific data collected, and of the relationships among these data, for the population sampled. A case study, on the other hand, justifies the selection of a particular case in terms of the goals of the study and existing theory and research, and needs a different kind of argument to support the generalizability of its conclusions (see Chapter 6).

Both approaches are legitimate in qualitative research. Interview studies, in particular, sometimes employ a "sampling" logic, selecting interviewees in order to generalize to some population of interest. In addition, the larger the study, the more feasible and appropriate a sampling approach becomes; large multisite studies in which generalizability is important (such as those described in Miles & Huberman, 1994) must pay considerable attention to issues of sampling and representativeness.

However, qualitative studies often employ small samples of uncertain representativeness, and this usually means that the study can provide only suggestive answers to any question framed in general terms, such as "How do kindergarten teachers assess the readiness of children for first grade?" A plausible answer to this generic question would normally require some sort of probability sampling from the population of all kindergarten teachers, and a larger sample than most qualitative studies can manage. Furthermore, the phrase "kindergarten teachers" is itself in need of further specification. Does it refer only to American teachers? Only to public school teachers? Only to certified teachers? These concerns, and analogous ones that could be raised about any research questions framed in generic terms, presuppose a "sample" framework for the study, and may push the study in a quantitative direction. (I say more about sampling in Chapter 5, and generalizability in Chapter 6.)

On the other hand, a qualitative study *can* confidently answer such a question posed in particularistic terms, such as "How do the kindergarten teachers *in this school* assess the readiness of children for first grade?" This way of stating the question, although it does not avoid issues of sampling, frames the study much more in "case" terms. The teachers are treated not as a *sample* from some much larger population of teachers to whom the study is intended to generalize, but as a *case* of a group of teachers who are studied in a particular context (the specific school and community). The *selection* of this particular case may involve considerations of representativeness (and certainly any attempt to generalize from the conclusions must take representativeness into account), but the primary concern of the study is not with generalization, but with developing an adequate description, interpretation, and explanation of this case.

INSTRUMENTALIST QUESTIONS AND REALIST QUESTIONS

As discussed in Chapter 2, social science was long dominated by the positivist view that only theoretical terms whose meaning could be precisely specified in terms of research operations and "objective" data (what came to be called "operational definitions") were legitimate in science. The statement that "Intelligence is whatever intelligence tests measure" is a classic example of this view. Although this position (often called "instrumentalism") has been abandoned by almost all philosophers of science, it still influences the way many researchers think about research questions. Advisors and reviewers often recommend framing research questions in terms of what the respondents say or report, or in terms of what can be directly observed, rather than in terms of inferred beliefs, behavior, or causal influences.

For example, Gail Lenehan, for her dissertation, proposed to interview nurses who specialize in treating sexual assault victims, focusing on their cognitive, behavioral, and emotional reactions to this work. Although there is considerable anecdotal evidence that these nurses often experience reactions similar to those of their victims, no one had systematically studied this phenomenon. Her research questions included the following:

1. What, if any, are the effects on nurses of working with rape victims?
2. Are there cognitive, psychological, and behavioral responses to having experiences of rape "shared" with them as well as witnessing victims' suffering after the assault?

Her proposal was not accepted, and the committee, in explaining its decision, argued (among other concerns) that

> the study relies solely on self-report data, but your questions do not reflect this limitation. Each question needs to be reframed in terms that reflect this limitation. Some examples might be: "How do nurses perceive and report . . . the effects of working with rape victims?" or "What specific cognitive, psychological (emotional?), and behavioral responses do nurses report?"

This disagreement illustrates the difference between instrumentalist and realist approaches (Norris, 1983) to research questions. Instrumentalists formulate their questions in terms of observable or measurable data. They worry about the validity threats (such as self-report bias) that the researcher risks in making inferences to unobservable phenomena, and prefer to stick with what they can directly verify. Realists, in contrast, do not assume that research questions and conclusions about feelings, beliefs, intentions, prior behavior, effects, and so on, need to be reduced to, or reframed as, questions and conclusions about the actual data that one collects. Instead, they treat these unobserved phenomena as *real,* and their data as *evidence* about these, to be used critically to develop and test ideas about the existence and nature of the phenomena (Campbell, 1988; Cook & Campbell, 1979; Maxwell, 1992, 2004a).

This is not a matter of philosophical hair-splitting; it has important implications for how you conduct the research, and each of the two approaches has its risks. The main risk of instrumentalist questions is that you will lose sight of what you're really interested in, and narrow your study in ways that exclude the actual phenomena you want to investigate, ending up with a rigorous but uninteresting conclusion. As in the joke about the man who was looking for his keys under the streetlight (rather than where he dropped them) because the light was better there, you may never find what you started out to look for. An instrumentalist approach to your research questions may also make it more difficult for you to address important goals of your study (such as developing programs to deal with the actual effects on nurses of talking to rape victims), and can inhibit your theorizing about phenomena that are not directly observable.

The main risk with realist questions, on the other hand, is that your increased reliance on inference may lead you to draw unwarranted conclusions, or to allow your assumptions or hopes to influence your results. My own preference is to use realist questions, and to address as systematically and rigorously as possible the validity threats that this approach involves. I have several reasons for this. First, the seriousness of these validity threats (such as self-report bias) depends on the topic, goals, and methods of the research, and thus needs to be assessed in the context of a particular study; these threats are often not as serious as instrumentalists imply. Second, there are usually effective ways to address these threats in a qualitative design; these will be discussed in Chapter 6. Finally, I take a realist position that unobservable phenomena (e.g., black holes, quarks, and the extinction of the dinosaurs) can be just as real as observable ones, and just as legitimate as objects of scientific study.

Thus, in my view, the risk of trivializing your study by restricting your questions to what can be directly observed is usually more serious than the risk of drawing invalid conclusions. What the statistician John Tukey said about precision is also true of certainty: "Far better an approximate answer to the right question, which is often vague, than an exact answer to the wrong question, which can always be made precise" (1962, p. 13; cited in Light & Pillemer, 1984, p. 105). My advice to students in Lenehan's position is to argue for the legitimacy of framing your questions in realist terms (which she successfully did). Even if you are required to restrict your *proposal* to instrumentalist questions, you should make sure that your actual *design* incorporates any realist concerns that you want your study to address.

One issue that is not entirely a matter of realism versus instrumentalism is whether research questions in interview studies should be framed in terms of the respondents' *perceptions* or *beliefs* about what happened, rather than what actually happened. This was an issue for Lenehan's study, described above; one recommendation of the committee was to focus the questions on how nurses *perceive* the effects of working with rape victims, rather than on the actual effects. Both of these are, in principle, realist questions, because, from a realist perspective, perceptions and beliefs are real phenomena, and neither is something that can be inferred with certainty from interview data. However, there is a definite instrumentalist bias in the reviewers' advice, because inferences to the actual situation or behavior being reported are in most circumstances more indirect and problematic than are inferences to the perspective of the respondent.

The decision between instrumentalist and realist questions should be based not simply on the seriousness of the risks and validity threats for each, but also on what you actually want to understand. In many qualitative studies, the real interest is in how participants make sense of what has happened (itself a real phenomenon), and how this perspective informs their actions, rather than in determining precisely what happened or what they did. Furthermore, in some circumstances you may be more interested in how participants organize and communicate their experiences (another real phenomenon) than in the "truth" of their statements (e.g., Linde, 1993). Jackson (1987, pp. 292–294), after finishing his study of death row inmates, was asked how he knew the men he interviewed were telling the truth, or even if they themselves believed what they told him. He eventually decided that he was in fact most interested in how the men constructed a presentation of self, a narrative of their life. As he said,

> Whether the condemned men who speak to you on these pages *believe* their presentations is interesting, but not finally important; what is important is first that they feel the need to organize their verbal presentations of themselves so they are rational, and second that they know how to do it. (p. 293)

VARIANCE QUESTIONS AND PROCESS QUESTIONS

Finally, I want to return to the distinction between variance theory and process theory that I introduced in Chapter 2, and relate this to the framing of research questions. Variance questions focus on difference and correlation; they often begin with "Does," "How much," "To what extent," and "Is there a relationship." Process questions, in contrast, focus on *how* things happen, rather than *whether* there is a particular relationship or how much it is explained by other variables. The fundamental distinction here is between questions that focus on variables and differences and those that focus on processes; it closely parallels the distinction between positivist and realist approaches to causation.

For example, asking "Do second-career teachers remain in teaching longer than teachers for whom teaching is their first career, and if so, what factors account for this?" is a variance question, because it implies a search for a difference and for the particular variables that explain the difference. An example of a process question would be "How do second-career teachers decide whether to remain in teaching or to leave?" The focus in the latter question is not in explaining a difference (a dependent variable) in terms of some independent variables, but on understanding how these teachers think about and make decisions on remaining in teaching.

In a qualitative study, it can be risky for you to frame your research questions in a way that focuses on differences and their explanation. This may lead you to begin thinking in variance terms, to try to identify the variables that will account for observed or hypothesized differences, and to overlook the real strength of a qualitative approach, which is in understanding the processes by which things take place. Variance questions are normally best answered by quantitative approaches, which are powerful ways of determining *whether* a particular result was related to one or another variable, and *to what extent* these are related. However, qualitative research is often better at showing *how* this occurred. (See the discussion of causality in Chapter 2.) Variance questions can be legitimate in qualitative research, but they are often best grounded in the answers to prior process questions.

Qualitative researchers thus tend to focus on three kinds of questions that are much better suited to process theory than to variance theory: (a) questions about the *meaning* of events and activities to the people involved in these, (b) questions about the influence of the physical and social *context* on these events and activities, and (c) questions about the *process* by which these events and activities and their outcomes occurred. (See the discussion of the goals of qualitative research in Chapter 2.) Because all of these types of questions involve situation-specific phenomena, they do not lend themselves to the kinds of comparison and control that variance theory requires. Instead, they generally involve an open-ended, inductive approach, in order to discover what these meanings and influences are and *how* they are involved in these events and activities—an inherently processual orientation.

One student, Bruce Wahl, wrote to me about having changed his research questions while he was analyzing the data for his dissertation, an evaluation of math projects for community college students that engaged different learning styles:

> I don't know if you remember, but two years ago when I was writing my proposal, you stressed that I should be writing my research questions beginning with words like "how" and "what" and "why" instead of the yes/no questions I was asking. For example, my first question was, "Do the projects help students to grasp mathematical concepts?" As I am writing up the interview results, I finally understand what you were saying. What I really wanted to know was "How do the projects help (or not help!) the students to grasp mathematical concepts?" It seems so clear now, it is a wonder that I didn't understand it back then. I have rewritten the 5 research questions for myself with that in mind and will include those new, and I hope, improved questions with the [dissertation] draft I deliver next week.

DEVELOPING RESEARCH QUESTIONS

Light et al. (1990) pointed out that formulating research questions is not a simple or straightforward task:

> Do not expect to sit down for an hour and produce an elaborate list of specific questions. Although you must take the time to do just that—sit down and write—your initial list will not be your final list. Expect to iterate. A good set of research questions will evolve, over time, after you have considered and reconsidered your broad research theme. (p. 19)

And they cautioned, "Be wary of the desire to push forward before going through this process" (p. 19).

What follows is an exercise for you to work through in developing your research questions. This exercise will not only generate research questions, but will also help you connect these questions to the other four components of your research design, in order to make these questions as relevant and practicable as possible. These connections are two-way streets; try to see not only what questions, or changes in questions, the other four components suggest, but also what changes in these other components your questions imply.

EXERCISE 4.1

Developing Your Research Questions

Like most of the other exercises in this book, this one asks you to write a memo that addresses the following sets of issues for your research. This involves trying to connect your tentative research questions to the other four components of your design. At this point, your answers to items 5 and 6 may need to be very tentative; that's fine. You can repeat this exercise as you get a better sense of what your study will look like.

1. Begin by setting aside whatever research questions you already have, and starting with your concept map (Chapter 3). What are the places in this map that you *don't* understand adequately, or where you need to test your ideas? Where are the holes in, or conflicts between, your experiential knowledge and existing theories, and what questions do these suggest? What could you learn in a research study that would help you to better understand what's going on with these phenomena? Try to write down all of the potential questions that you can generate from the map.

2. Next, take your original research questions and compare them to the map and the questions you generated from it. What would answering these questions tell you that you *don't* already know? What changes or additions to your questions does your map suggest? Conversely, are there places where your questions imply things that *should* be in your map, but aren't? What changes do you need to make to your map?

3. Now go through the same process with your researcher identity memo (Chapter 2). What could you learn in a research study that would help to accomplish these goals? What questions does this imply? Conversely, how do your original questions connect to your reasons for conducting the study? How will answering these *specific* questions help you to achieve your goals? Which questions are most *interesting* to you, personally, practically, or intellectually?

4. Now *focus*. What questions are most *central* for your study? How do these questions form a coherent set that will guide your study? You can't study everything interesting about your topic; start making choices. Three or four main questions are usually a reasonable maximum for a qualitative study, although you can have additional subquestions for each of the main questions.

5. In addition, you need to consider what sort of study could actually answer the questions you pose. Connect your questions to the methods you might use: *Could* your questions be answered by these methods and the data that they would provide? What methods would you *need* to use to collect data that would answer these questions? Conversely, what questions can a qualitative study of the kind you are planning productively address? At this point in your planning, this may primarily involve

"thought experiments" about the way you will conduct the study, the kinds of data you will collect, and the analyses you will perform on these data. This part of the exercise is one you can usefully repeat when you have developed your methods and validity concerns in more detail; Exercise 5.2, in the next chapter, also addresses these issues.

6. Assess the potential answers to your questions in terms of validity. What are the plausible validity threats and alternative explanations that you would have to rule out? How might you be wrong, and what implications does this have for the way you frame your questions?

Don't get stuck on trying to precisely frame your research questions, or in specifying in detail how to measure things or gain access to data that would answer your questions. Try to develop some meaningful and important questions that would be *worth* answering. Feasibility is obviously an important issue in doing research, but focusing on it at the beginning can abort a potentially valuable study. My experience is that there are very few important questions that can't be potentially answered by some sort of research.

An extremely valuable additional step is to share your questions and your reflections on these with a small group of fellow students or colleagues. Ask them if they understand the questions and why these would be worth answering, what other questions or changes in the questions they would suggest, and what problems they see in trying to answer them. If possible, tape-record the discussion; afterward, listen to the tape and take notes.

5

Methods

What Will You Actually Do?

In this chapter, I discuss some of the key issues involved in planning what you will do in conducting your research. These issues are not limited to qualitative data collection (primarily participant observation and interviewing), but also include establishing research relationships with those you study, selecting sites and participants, and analyzing the data that you collect. The focus is on how to *design* the use of specific methods in a qualitative study, not on how to actually do qualitative research; I am assuming that you already know (or are learning) the latter.

At the outset, I want to emphasize that there is no "cookbook" for doing qualitative research. The appropriate answer to almost any question about the use of qualitative methods is "it depends." Decisions about research methods depend on the specific context and issues you are studying, as well as on other components of your design. The bottom line for any decision is the actual consequences of using it in your study; what would be an excellent decision in one study could be a disaster in another. What I want to discuss here are some of the things that your methodological decisions depend *on*—the issues that you will need to think about in designing your research methods.

In addition, the "data" in a qualitative study can include virtually anything that you see, hear, or that is otherwise communicated to you while conducting the study; there is no such thing as "inadmissible evidence" in trying to understand the issues or situations you are studying. (However, there may be evidence that you are ethically prohibited from *citing* in what you write, if it could violate confidentiality or privacy or be potentially damaging to particular individuals.) Qualitative data are not restricted to the results of specified "methods"; as noted earlier, you *are* the research instrument in a qualitative study, and your eyes and ears are the tools you use to make sense of what is going on. In planning your research methods, you should always include whatever informal data-gathering strategies are feasible, including "hanging out," casual conversations, and incidental observations. This is particularly important in an interview study, where such information can provide important contextual information, a different perspective from the interviews, and a check on your interview data. As Dexter (1970) emphasized,

> no one should plan or finance an entire study in advance with the expectation of relying chiefly upon interviews for data unless the interviewers have enough relevant background to be sure that they can make sense out of interview conversations or unless there is a reasonable hope of being able to hang around or in some way observe so as to learn what it is meaningful and significant to ask. (p. 17)

Such data should be systematically recorded in memos or a field journal.

STRUCTURED AND UNSTRUCTURED APPROACHES

One important issue in designing a qualitative study is to what extent you should decide on your methods in advance, rather than developing or modifying these during the research. Some qualitative researchers believe that, because qualitative research is necessarily inductive and "grounded," any substantial prior structuring of the methods leads to a lack of flexibility to respond to emergent insights, and can create methodological "tunnel vision" in making sense of your data. This decision is often justified on philosophical or ethical grounds as well; structured approaches are identified with quantitative research, positivism, or power inequalities between researcher and researched. The choice between structured and unstructured methods is rarely discussed in a way that clarifies the relative advantages and disadvantages of each. (Significant exceptions are Miles & Huberman, 1994; Robson, 2002; Sayer, 1992.)

Structured approaches can help to ensure the comparability of data across individuals, times, settings, and researchers, and are thus particularly useful in answering variance questions, questions that deal with *differences* between things. Unstructured approaches, in contrast, allow you to focus on the *particular* phenomena being studied, which may differ from others and require individually tailored methods. They trade generalizability and comparability for internal validity and contextual understanding, and are particularly useful in revealing the processes that led to specific outcomes, what Miles and Huberman (1994) called "local causality" (cf. Maxwell, 2004a).

However, Miles and Huberman also cautioned that

> Highly inductive, loosely designed studies make good sense when experienced researchers have plenty of time and are exploring exotic cultures, understudied phenomena, or very complex social phenomena. But if you're new to qualitative

> studies and are looking at a better understood phenomenon within a familiar culture or subculture, a loose, inductive design is a waste of time. Months of field-work and voluminous case studies may yield only a few banalities. (1994, p. 17)

They also pointed out that prestructuring your methods reduces the amount of data that you have to deal with, simplifying the analytic work required (1994, p. 16).

In general, I agree with Miles and Huberman's assessment, although I think their involvement with multiple-site research has led them to advocate more prestructuring than is appropriate for most single-site studies. However, like nearly everyone else, they treat prestructuring as a single dimension, and view it in terms of metaphors such as hard versus soft and tight versus loose. Such metaphors, in addition to their one-dimensional implications, have value connotations (although these are different for different people) that can interfere with your assessment of the tradeoffs involved in particular design decisions and the best ways to combine different aspects of prestructuring within a single design. These metaphors can lead you to overlook or ignore the numerous ways in which studies can vary, not just in the *amount* of prestructuring, but in *how* prestructuring is used.[1]

For example, Festinger, Riecker, and Schachter (1956), in a classic study of an end-of-the-world cult, employed an extremely open approach to data collection, relying primarily on descriptive field notes from covert participant observation in the cult. However, they used these data for a confirmatory test of explicit hypotheses based on a prior theory, rather than to inductively develop new questions or theory (cf. Maxwell & Loomis, 2002, pp. 260–263). In contrast, the approach often known as ethnoscience or cognitive anthropology (Spradley, 1979; Werner & Schoepfle, 1987) employs highly structured data collection techniques, but interprets these data in a largely inductive manner, with very few preestablished categories. Thus, the decision you face is not primarily *whether* or *to what extent* you prestructure your study, but *in what ways* you do this, and *why*.

Finally, it is worth keeping in mind that you can lay out a *tentative* plan for some aspects of your study in considerable detail, but leave open the possibility of substantially revising this if necessary. (See the evolution of Maria Broderick's research design, presented in Example 1.1.) The degree to which you prestructure your anticipated research methods is a separate decision from how much flexibility you leave yourself to revise the plan during your study. Emergent insights may require new selection plans, different kinds of data, and different analytic strategies. As stated earlier, all research has an implicit, if not explicit, design. Avoiding decisions about your design may mean only that you aren't examining the design that is implicit in your thinking, and are failing to recognize the consequences that these implicit decisions will have. Deliberate attention to these consequences can help you to construct a design that will enable you to answer your questions, advance your goals, and possibly save you a lot of trouble.

I see qualitative methods—what you will actually do in conducting a qualitative study—as having four main components:

1. The research relationships that you establish with those you study
2. Site and participant selection: what settings or individuals you select to observe or interview, and what other sources of information you decide to use
3. Data collection: how you gather the information you will use
4. Data analysis: what you do with this information in order to make sense of it

This is a somewhat broader definition of "methods" than is usual in discussions of research design. My justification for this definition is that all of these components are important aspects of how you conduct your study, and affect the value and validity of your conclusions. It is therefore useful to think about these as *design* decisions—key issues that you should consider in planning your study, and that you should rethink as you are engaged in it. In the rest of this chapter, I will discuss what I see as the most important considerations that should affect your decisions about each of these components.

NEGOTIATING RESEARCH RELATIONSHIPS

The relationships that you create with participants in your study (and also with others, sometimes called "gatekeepers," who can facilitate or interfere with your study) are an essential part of your methods, and how you initiate and negotiate these relationships is a key *design* decision. Bosk (1979, p. ix) noted that fieldwork is a "body-contact" sport, and your research relationships create and structure this contact. Conversely, your ongoing contact with participants continually restructures these relationships. These are both aspects of what Hammersley and Atkinson (1995, p. 16) called "reflexivity"—the fact that the researcher is part of the social world he or she studies.

In many books on qualitative methods, these relationships are conceptualized as "gaining access" to the setting (e.g., Bogdan & Biklen, 2003, pp. 75–80; Glesne, 1999, pp. 39–40) or as "negotiating entry" (e.g., Marshall & Rossman, 1999, p. 82). While this is one important *goal* in negotiating a relationship, such phrases may lead you to think that this is something that, once "achieved," requires no further attention. The process of negotiating a relationship is much more complex than these phrases suggest; not only does it typically require ongoing negotiation and renegotiation of your relationships with those you study, but it rarely involves any approximation to total access. Nor is total access usually necessary for a successful study; what you need are relationships that allow you to ethically gain the information that can answer your research questions.

Conceptualizing your relationships in terms of "rapport" is also problematic, because it represents a relationship by a single continuous variable, rather than emphasizing the *nature* of that relationship. Seidman (1998, pp. 80–82) made the important point that it is possible to have too much rapport, as well as too little, but I would add that it is the *kind* of rapport, as well as the amount, that is critical. A participant can be very engaged intellectually in an interview, but not be revealing anything deeply personal, and for some studies this kind of relationship may be ideal. Conversely, someone may be very open about personal matters to a stranger whom they never expect to see again, but not be willing to engage in any critical reflection on this material.

Thus, the relationship you have with a participant in your study is a complex and changing entity. In qualitative studies, the researcher is the instrument of the research, and the research relationships are the means by which the research gets done. These relationships have an effect not only on the participants in your study, but also on you, as both researcher and human being, as well as on other parts of the research design. In particular, the research relationships you establish can facilitate or hinder other components of the research design, such as participant selection and data collection. For example, in my dissertation research in an Inuit community, I made arrangements to live with different families on a monthly basis. This gave me access to detailed information about a wider range of families than is often available to anthropologists, who typically establish close ties with a small number of individuals or families. However, the ways in which this arrangement was negotiated made it difficult for me to develop working relationships with those families with whom I did not live (Maxwell, 1986). Rabinow (1977) provided an insightful account of the way in which his changing relationships with his Moroccan informants affected his research plans, and Bosk (1979) explained how his relationships with the surgeons he studied both facilitated and constrained his research. Many other accounts by qualitative researchers of their research provide similar insights; rather than attempting to sum these up in a few, only partially generalizable "guidelines," I urge you to read widely in the literature on this topic, so that your decisions can be informed by a range of other researchers' experiences.

I want to emphasize that these are *design* decisions, not simply "practical" issues that are separate from design. You will need to reflect on the particular decisions (conscious or unconscious) that you make about your relationships, as well as on the relationship issues that you will face in doing the study, and the effects these can have on your research. How to make these decisions gets deeper into issues of qualitative methods than this book can go, but the principle stated by Weiss (1994) for interview studies is valid for qualitative research in general:

> What is essential in interviewing is to maintain a working research partnership. You can get away with phrasing questions awkwardly and with a variety of other errors that will make you wince when you listen to the tape later. What you can't get away with is failure to work with the respondent as a partner in the production of useful material. (p. 119)

In addition to these considerations, there are philosophical, ethical, and political issues that should inform the kinds of relationships that you want to establish. In recent years, the dominance of the traditional research relationship has been challenged by alternative modes of research that involve quite different sorts of relationships between the researcher and the researched, and in some cases break down this distinction entirely. For example, Tolman and Brydon-Miller (2001) advocated "interpretive and participatory action methods" in qualitative research, methods that are "relational in that they acknowledge and actively involve the relationships between researchers and participants, as well as their respective subjectivities" (p. 5). They believed that qualitative research should be "participatory" in the sense of working collaboratively with research participants to generate knowledge that is useful to the participants as well as to the researcher, contributing to personal and social transformation (pp. 3–4). Similarly, Lawrence-Lightfoot and Davis (1997) criticized the tendency, even in qualitative research, to treat relationships as a tool or strategy for gaining access to data, rather than as a connection (p. 135). They argued that "relationships that are complex, fluid, symmetric, and reciprocal—that are shaped by both researchers and actors—reflect a more responsible ethical stance *and* are likely to yield deeper data and better social science" (pp. 137–138), and they emphasized the continual creation and renegotiation of trust, intimacy, and reciprocity.

Burman (2001) cautioned, however, that the dominant humanitarian/democratic agenda of qualitative research, as well as particular goals such as "relationship," "equality," and "participation," are easily co-opted into the perpetuation of existing power relationships, and she asserted that "the progressive . . . character of research is always ultimately a matter of politics, not technique" (pp. 270–271). My advocacy of incorporating research relationships into your research design is not an advocacy of any *particular* type of relationship. Although I mostly agree with Weiss, Tolman and Brydon-Miller, and Lawrence-Lightfoot and Davis, the types of relationships (and goals) that are ethically and politically appropriate depend on the particular context (including the participants' views), and should always be subjected to the sort of critique raised by Burman.

Whatever your methodological and political views, remember that what is a "research project" for you is always, to some degree, an intrusion into the lives of the participants in your study. You need to follow the rules for considerate interaction with others, and to *learn* these rules if, for the people or setting you're studying, they are different from what you're used to. A basic strategy to use here is to put yourself in their position, and ask how you would feel if someone did to you what you are thinking of doing, making allowances for differences in culture and norms. As Eeyore said, "A little consideration, a little thought for others, makes all the difference."

Finally, think about what you can give to participants in return for the time and inconvenience of being involved in your research. What can you do to make people feel that this has been a worthwhile experience and that they aren't just being "used"? This can range from helping out in the setting you're studying, to providing some gift or service, to simply being an empathic listener. What it's appropriate to offer depends on the setting and individual and on what you ask that person to do, but *some* acknowledgment of your appreciation is almost always required. As one of my students, Caroline Linse, reminded me, "The interview isn't over until the thank-you note is delivered."

EXAMPLE 5.1

Negotiating Relationships in a Practitioner Research Study

Bobby Starnes, a doctoral student with extensive experience as a teacher and administrator and a longtime political commitment to collaborative decision making, came to the Harvard Graduate School of Education in order to see how what she had learned about teaching and learning with children could inform her work with adults. When she was seeking a dissertation study that would allow her to apply and test her ideas, she was hired as director of a daycare center, serving a low-income population, which had a history of ineffective, top-down management. Her dissertation research was a study of what happened when she attempted to implement a system of shared decision making in this setting—how the system evolved, and how it affected staff morale, competence, and performance.

Bobby's study required her to have a very different relationship to her study participants than that found in most research; she was both their boss and a researcher trying to understand their perspective on the organizational changes she instituted. In addition, her political views led her to design a study in which she was engaged in real-world action to improve people's lives, not ivory-tower research. This combination posed both substantial risks of bias and distortion of the data, and unique opportunities for understanding the process of organizational change. It was thus absolutely essential for her study that her participants be open about their perceptions and feelings, and that they trust her not to use the data she collected in ways that would be harmful to them.

Bobby was able to accomplish this by establishing an organizational climate in which staff were not afraid to voice their opinions and disagree with her, and in which they were convinced that she would not violate confidences or take action against them as a result of what she learned. (Obviously, this was not an easy task, and required all of her skill and experience to carry out; for a detailed description of how she did this, see Starnes, 1990.) Without this relationship, the conclusions of her study would not have been trustworthy. However, she did not assume that the relationship that she had with her staff would automatically eliminate problems of distortion and concealment. She gathered some data by anonymous questionnaires, and had another researcher conduct half of the final interviews.

Because the impact of these issues is particular to each individual study, the best strategy for dealing with them is to think about them in the context of your own research. The following exercise should help you to do this.

EXERCISE 5.1

Reflecting on Your Research Relationships

This exercise involves writing a memo reflecting on your relationships (actual or planned) with participants and other important people you plan to involve in your research, how you will present yourself and your research, and what arrangements you expect to negotiate for doing the research and reporting your results. The following questions are ones you should keep in mind as you work on this memo:

1. What kinds of relationships have you established, or plan to establish, with the people in your study or setting? How did these relationships develop, or how do you plan to initiate and negotiate them? *Why* have you planned to do this? *How* do you think these relationships could (or already have) facilitate or impede your study? What alternative kinds of relationships could you create, and what advantages and disadvantages would these have?
2. How do you think you will be seen by the people you interact with in your research? How will this affect your relationships with these people? What could you do to better understand and (if necessary) modify this perception?
3. What explicit agreements do you plan to negotiate with the participants in your study about how the research will be conducted and how you will report the results? What *implicit* understandings about these issues do you think these people (and you) will have? How will both the implicit and the explicit terms of the study affect your relationships and your research? Do any of these need to be discussed or changed?
4. What ethical issues or problems do these considerations raise? How do you plan to deal with these?

As with the memo on research questions (Exercise 4.1), this can be a valuable memo to discuss with colleagues or fellow students.

SITE AND PARTICIPANT SELECTION

Decisions about where to conduct your research and whom to include (what is traditionally called "sampling") are an essential part of your research methods. Even a single case study involves a choice of this case rather than others, as well as requiring such decisions *within* the case itself. Miles and Huberman asked, "Knowing, then, that one cannot study everyone everywhere doing everything, even within a single case, how does one limit the parameters of a study?" (1984, p. 36). They argued that

> Just *thinking* in sampling-frame terms is healthy methodological medicine. If you are talking with one kind of informant, you need to consider *why* this kind of informant is important, and, from there, which *other* people should be interviewed. This is a good, bias-controlling exercise.
>
> Remember that you are not only sampling *people,* but also *settings, events, and processes.* It is important to line up these parameters with the research questions as well, and to consider whether your choices are doing a representative, time-efficient job of answering them. The settings, events, or processes that come rapidly to mind at the start of the study may not be the most pertinent or data-rich ones. A systematic review can sharpen early and later choices. (1984, p. 41)

Miles and Huberman (1994, pp. 27–34) and LeCompte and Preissle (1993, pp. 56–85) provided valuable discussions of the whole issue of sampling decisions, and I will not repeat all of their advice here. Instead, I want to talk about the *purposes* of different selection strategies, and some of the considerations that are relevant to such decisions.

First, the term "sampling" is problematic for qualitative research, because it implies the purpose of "representing" the population sampled. Quantitative methods texts typically recognize only two main types of sampling: probability sampling (such as random sampling) and convenience sampling (e.g., Light, Singer, & Willett, 1990, pp. 56–57). In probability sampling, each member of the population has a known, nonzero probability of being chosen, allowing statistical generalization from the sample to the population of interest. Light et al. stated that "probability samples are a paragon of high-quality research" (p. 56), a view that is widespread. As a result, any nonprobability sampling strategy is seen as "convenience sampling," and is strongly discouraged.

This view ignores the fact that, in qualitative research, the typical way of selecting settings and individuals is neither probability sampling nor convenience sampling. It falls into a third category, which I will call *purposeful selection* (Light et al., 1990, p. 53); other terms are *purposeful sampling* (Patton, 1990, p. 169) and *criterion-based selection* (LeCompte & Preissle, 1993, p. 69). This is a strategy in which particular settings, persons, or activities are selected deliberately in order to provide information that can't be gotten as well from other choices. For example, Weiss argued that many qualitative interview studies do not use "samples" at all, but *panels*—"people who are uniquely able to be informative because they are expert in an area or were privileged witnesses to an event" (1994, p. 17); this is one form of purposeful selection. Selecting those times, settings, and individuals that can provide you with the information that you need in order to answer your research questions is the most important consideration in qualitative selection decisions.

Patton (1990, pp. 169–186) and Miles and Huberman (1994, pp. 27–29) described a large number of types of sampling that can be employed in qualitative research, almost all of which are forms of purposeful selection. Patton mentioned "convenience sampling" only to warn against its use, claiming that

> while convenience and cost are real considerations, they should be the last factors to be taken into account after strategically deliberating on how to get the most information of the greatest utility from the limited number of cases to be sampled. . . . *Convenience sampling is neither purposeful nor strategic.* (p. 181; emphasis in original)

However, Weiss (1994, pp. 24–29) argued that there are situations in which convenience sampling is the only feasible way to proceed—for example, in attempting to learn about a group that is difficult to gain access to, or a category of people who are relatively rare in the population and for whom no data on membership exist, such as "house husbands." He listed several strategies for maximizing the value of such convenience samples.[2]

In qualitative studies with large sample sizes (e.g., Huberman, 1989) in which generalizability is an important goal, random sampling is a valid and often appropriate procedure. However, simple random sampling is a very poor way to draw a small sample, due to the high likelihood of substantial chance variation in such samples. Most of the advantages of random sampling depend on having a reasonably large sample size to make such variations unlikely. Light et al., in discussing *site* selection, stated that "with only a limited number of sites, consider *purposeful selection,* rather than relying on the idiosyncrasies of chance" (1990, p. 53); the same logic applies to selecting interview participants and observation settings.

There are a few circumstances in which random sampling can be useful in a small-scale qualitative study. Bobby Starnes, in her study of shared decision making in a daycare center (Example 5.1), used stratified random sampling of center staff when she had more volunteers than she could interview, mainly in order to avoid the perception of favoritism in selecting interviewees. However, in one case she altered the random selection in order to include a point of view that she believed would not otherwise have been represented (Starnes, 1990, p. 33).

There are at least four possible goals for purposeful selection; Creswell (2002, pp. 194–196) listed others, but I see these four as most important. The first is achieving representativeness or typicality of the settings, individuals, or activities selected. Because, as noted previously, random sampling is likely to achieve this only with a large sample size, deliberately selecting cases, individuals, or situations that are known to be typical provides far more confidence that the conclusions adequately represent the average members of the population than does a sample of the same size that incorporates substantial random or accidental variation.

The second goal that purposeful selection can achieve is the opposite of the first—to adequately capture the heterogeneity in the population. The purpose here is to ensure that the conclusions adequately represent the entire *range* of variation, rather than only the typical members or some "average" subset of this range; Guba and Lincoln (1989, p. 178) referred to this as "maximum variation" sampling. This is best done by defining the dimensions of variation in the population that are most relevant to your study and systematically selecting individuals or settings that represent the most important possible variations on these dimensions.[3] The tradeoff between this approach and selecting a more homogeneous sample is that you have less data about any *particular* kind of case, setting, or individual within the study, and will not be able to say as much in depth about typical instances.

The third possible goal is to deliberately examine cases that are critical for the theories that you began the study with, or that you have subsequently developed.[4] Extreme cases often provide a crucial test of these theories, and can illuminate what is going on in a way that representative cases cannot. For example, Wievorka (1992) described a study in which the researcher, in order to test the view that the working class was not being assimilated into middle-class society, selected a case that would be highly unfavorable to this position: workers who were extremely affluent. The finding that these workers still retained a clear working-class identity provided far more convincing support for his conclusions than a study of "typical" workers would.

A fourth goal in purposeful selection can be to establish particular comparisons to illuminate the reasons for differences between settings or individuals. While such comparisons are less common in qualitative than in quantitative research, comparative designs are often used in multicase qualitative studies, as well as in mixed-method research (Maxwell & Loomis, 2002). However, explicit comparisons are usually not very productive in a small-scale qualitative study, because the small number of cases in any group limits your ability to draw firm conclusions about the differences between the groups. In addition, an emphasis on comparisons can skew your study toward the analysis of differences (variance theory), as described in Chapter 4, and lead you to neglect the main strength of qualitative research, which is its ability to elucidate *local* processes, meanings, and contextual influences in particular settings or cases.

In many situations, selection decisions require considerable knowledge of the setting of the study. In Jane Margolis's study of classroom discourse norms in a college department (1990), she could interview only a small percentage of the students, and needed to develop some criteria for selecting participants. Her committee (of which I was a member) recommended that she interview sophomores and seniors, believing that this would provide the optimal diversity of views. When she consulted with members of the department, however, they told her that sophomores were too new to the department to fully understand the norms of discourse, while seniors were too deeply involved in their theses and in planning for graduation to be good informants. Juniors turned out to be the only appropriate choice.

Selection decisions should also take into account the feasibility of access and data collection, your research relationships with study participants, validity concerns, and ethics. For example, in Martha Regan-Smith's study of how medical school teachers help students learn basic science (see the Appendix), her choice of four award-winning teachers was based not only on the fact that these teachers were the most likely to exhibit the phenomena she was interested in (purposeful selection), but also because (as a fellow award winner) she had a close and collegial relationship with them that would facilitate the study. In addition, as exemplary teachers, they would be more likely to be candid about their teaching, and the research would be less likely to create ethical problems arising from her discovery of potentially damaging information about them.

One particular selection problem in qualitative studies has been called "key informant bias" (Pelto & Pelto, 1975, p. 7). Qualitative researchers sometimes rely on a small number of informants for a major part of their data, and even when these informants are purposefully selected and the data themselves seem valid, there is no guarantee that these informants' views are typical. In addition, Poggie (1972) presented evidence that key informants themselves assume greater uniformity than actually exists. There is increasing recognition that cultural groups incorporate substantial diversity and that homogeneity cannot be assumed (Hannerz, 1992; Maxwell, 1995, 1999). Thus, you need to do systematic sampling in order to be able to claim that key informants' statements are representative of the group as a whole (Heider, 1972; Sankoff, 1971).

DECISIONS ABOUT DATA COLLECTION

Most qualitative methods texts devote considerable space to the strengths and limitations of different qualitative data collection methods (see particularly Bogdan & Biklen, 2003; Patton, 2001), and I don't want to repeat these discussions here. Instead, I want to address two key conceptual issues in selecting and using different data collection methods: the relationship between research questions and data collection methods and the triangulation of different methods. (The relative advantages of structured and unstructured methods, discussed previously, are also important considerations in planning data collection methods.)

The Relationship Between Research Questions and Data Collection Methods

The point that I want to emphasize here is that the methods you use to collect your data (including your interview questions) don't necessarily resemble, or follow by logical deduction from, the research questions; the two are distinct and separate parts of your design. This can be a source of confusion, because researchers often talk about "operationalizing" their research questions, or of "translating" the research questions into interview questions. Such language is a vestige of logical positivist views of the relationship between theory and data, views that have been almost completely abandoned by philosophers (Phillips, 1987). There is no way to mechanically "convert" research questions into methods; your methods are the *means* to answering your research questions, not a logical transformation of the latter. Their selection depends not only on your research questions, but also on the actual research situation and on what will work most effectively in that situation to give you the data you need.

A striking example of this, concerning interview questions, was provided by Kirk and Miller (1986, pp. 25–26), who conducted research in Peru on the use of coca leaves. Their open-ended questions about coca use, drawn fairly directly from their research questions, elicited a uniform, limited set of beliefs and practices that simply confirmed the things they had already read about coca. Frustrated and getting desperate, they began asking less logical questions, such as, "When do you give coca to animals?" or "How did you discover that you didn't like coca?" Taken off guard, their informants began to open up and talk about their personal experience with coca, which was far more extensive than the previous data would have indicated.

This is an extreme case, but it holds in principle for any study. Your research questions formulate what you want to understand; your *interview* questions are what you ask people in order to gain that understanding. The development of good interview questions (and observational strategies) requires creativity and insight, rather than a mechanical conversion of the research questions into an interview guide or observation schedule, and depends fundamentally on how the interview questions and observational strategies will actually work in practice.

This doesn't mean that you should conceal your research questions from participants, or treat them simply as subjects to be manipulated to produce the data you need, as discussed previously in the section titled "Negotiating Research Relationships." Carol Gilligan (personal communication) emphasized the value of asking your interviewees "real questions," ones to which you are genuinely interested in the answer, rather than contrived questions designed to elicit particular sorts of data. Doing this creates a more symmetrical and collaborative relationship in which participants are able to bring their own knowledge to bear on the questions in ways that you might never have anticipated.

There are two important implications that the lack of a direct logical connection between research questions and interview questions has for your research. First, you need to anticipate, as best you can, how particular questions will actually work in practice—how people will understand them, and how they are likely to respond. Try to put yourself in your interviewee's place and imagine how you would react to these questions (this is another use of "thought experiments") and get feedback from others on how they think the questions (and the interview guide as a whole) will work. Second, if at all possible, you should *pilot-test* your interview guide with people as much like your planned interviewees as possible, to determine if the questions work as intended and what revisions you may need to make (see Example 3.4).

In addition, there are some cultures, settings, and relationships in which it is not appropriate or productive to conduct interviews, or even to ask questions, as a way of gaining information. C. Briggs (1986) described how, in his research on traditional religious wood carving in a Spanish-speaking community in northern New Mexico, the cultural norms of the community made the interviews he had planned to conduct completely inappropriate, and rendered these largely useless when he persisted with them. This situation forced him to discover the culturally appropriate way to learn about this topic, which was by apprenticeship. Similarly, Mike Agar, conducting research on heroin use, found that, on the streets, you don't ask questions. First, doing so raises suspicions that you will pass information on to the police or use it to cheat or rob the person you asked. Second, asking questions shows that you're not "hip," and therefore don't belong there (Hammersley & Atkinson, 1995, p. 128). Hammersley and Atkinson (1995) provided other examples of how traditional interviews may be inappropriate or unproductive (pp. 127–130), and C. Briggs (1986) showed how interviewing imposes particular Anglo-American discourse norms on one's participants, which can damage the relationship or reduce the amount of useful information you get.

This lack of a deductive relationship between questions and methods also holds, more obviously, for observation and other data collection methods. As with interviews, you need to anticipate what information you will actually be able to collect, in the setting studied, using particular observational or other methods, and, if possible, you should pretest these methods to determine if they will actually provide this information. Your data collection strategies will probably go through a period of focusing and revision, even in a carefully designed study, to enable them to better provide the data that you need to answer your research questions and to address any plausible validity threats to these answers.

Triangulation of Data Collection Methods

Collecting information using a variety of sources and methods is one aspect of what is called *triangulation* (Fielding & Fielding, 1986). This strategy reduces the risk that your conclusions will reflect only the systematic biases or limitations of a specific source or method, and allows you to gain a broader and more secure understanding of the issues you are investigating. I discuss the use of triangulation generally, as a way to deal with validity threats, in Chapter 6; here, I want to focus specifically on combining different data collection methods.

Bobby Starnes's study (Example 5.1) provides a good illustration of the use of triangulation. She used four sources of data (the direct-care staff, her administrative team, her own notes and journals, and the center records) and several different methods of collecting these data. For example, the data from the staff were collected through journals, formal and informal interviews, participation in center activities, and anonymous questionnaires. These multiple sources and methods give her conclusions far more credibility than if she had been limited to one source or method.

One belief that inhibits triangulation is the widespread (though often implicit) assumption that observation is mainly useful for describing behavior and events, while interviewing is mainly useful for obtaining the perspectives of actors. It is true that the *immediate* result of observation is description, but this is equally true of interviewing: The latter gives you a description of what the informant *said,* not a direct understanding of his or her perspective. Generating an interpretation of someone's perspective is inherently a matter of inference from descriptions of that person's behavior (including verbal behavior), whether the data are derived from observations, interviews, or some other source such as written documents (Maxwell, 1992).

While interviewing is often an efficient and valid way of understanding someone's perspective, observation can enable you to draw inferences about this perspective that you couldn't obtain by relying exclusively on interview data. This is particularly important for getting at tacit understandings and "theory-in-use," as well as aspects of the participants' perspective that they are reluctant to directly state in interviews. For example, watching how a teacher responds to boys' and girls' questions in a science class may provide a much better understanding of the teacher's actual views about gender and science than what the teacher says in an interview.

Conversely, although observation often provides a direct and powerful way of learning about people's behavior and the context in which this occurs, interviewing can also be a valuable way of gaining a description of actions and events—often the *only* way, for events that took place in the past or ones to which you cannot gain observational access. Interviews can provide additional information that was missed in observation, and can be used to check the accuracy of the observations. However, in order for interviewing to be useful for this purpose, you need to ask about *specific* events and actions, rather than posing questions that elicit only generalizations or abstract opinions (Weiss, 1994, pp. 72–76). In both of these situations, triangulation of observations and interviews can provide a more complete and accurate account than either could alone.

DECISIONS ABOUT DATA ANALYSIS

Analysis is often conceptually separated from design, especially by writers who see design as what happens *before* the data are actually collected. Here, I treat analysis as a part of design (Coffey & Atkinson, 1996, p. 6), and as something that must itself be designed. Any qualitative study requires decisions about how the analysis will be done, and these decisions should inform, and be informed by, the rest of the design. The discussion of data analysis is often the weakest part of a qualitative proposal; in extreme cases, it consists entirely of generalities and "boilerplate" language taken from methods texts, and gives little sense of how the analysis will actually be done.

One of the most common problems in qualitative studies is letting your unanalyzed field notes and transcripts pile up, making the task of final analysis much more difficult and discouraging. There is a mountaineer's adage that the experienced climber begins lunch immediately after finishing breakfast, and continues eating lunch as long as he or she is awake, stopping briefly to eat supper (Manning, 1960, p. 54). In the same way, the experienced qualitative researcher begins data analysis immediately after finishing the first interview or observation, and continues to analyze the data as long as he or she is working on the research, stopping briefly to write reports and papers. Heinrich's (1984) rationale for immediately analyzing his biological data applies equally to the social sciences:

> On a research project I usually try to graph my data on the same day I collect them. From day to day the points on the graph tell me about my progress. It's like a fox pursuing a hare. The graph is the hare's track, and I must stay close to that hare. I have to be able to react and change course frequently. (p. 71)

As Coffey and Atkinson (1996) stated, "We should never collect data without substantial analysis going on simultaneously" (p. 2). Again, this is a *design* decision, and how it will be done should be systematically planned (and explained in your proposal).

STRATEGIES FOR QUALITATIVE DATA ANALYSIS

For novices, data analysis is probably the most mysterious aspect of qualitative research. As with data collection methods, the following discussion is not intended to explain how to *do* qualitative data analysis; some good sources for this are Bogdan and Biklen (2003, chap. 5), Coffey and Atkinson (1996), Emerson, Fretz, and Shaw (1995), Miles and Huberman (1994), Strauss and

Corbin (1990), and Weiss (1994, chap. 6). Instead, I want to provide an overview of the different strategies and conceptual tools for qualitative analysis, and then discuss some specific issues in making decisions about analytic methods.

The initial step in qualitative analysis is *reading* the interview transcripts, observational notes, or documents that are to be analyzed (Emerson et al., 1995, pp. 142–143). Listening to interview tapes prior to transcription is also an opportunity for analysis, as is the actual process of transcribing interviews or of rewriting and reorganizing your rough observation notes. During this reading or listening, you should write notes and memos on what you see or hear in your data, and develop tentative ideas about categories and relationships.

At this point, you have a number of analytic options. These fall into three main groups: (1) memos, (2) categorizing strategies (such as coding and thematic analysis), and (3) connecting strategies (such as narrative analysis) (Maxwell & Miller, n.d.). Unfortunately, many texts and published articles deal explicitly only with coding, giving the impression that coding *is* qualitative data analysis. In fact, most researchers informally use other strategies as well; they just don't describe these as part of their analysis. I want to emphasize that reading and thinking about your interview transcripts and observation notes, writing memos, developing coding categories and applying these to your data, and analyzing narrative structure and contextual relationships are *all* important types of data analyses. Their use needs to be planned (and carried out) in order to answer your research questions and address validity threats.

As discussed in Chapter 1, memos can perform other functions not related to data analysis, such as reflection on methods, theory, or purposes. However, they are also an essential technique for qualitative analysis (Miles & Huberman, 1994, pp. 72–75; Strauss & Corbin, 1990, pp. 197–223). You should regularly write memos while you are doing data analysis; memos not only capture your analytic thinking about your data, but also *facilitate* such thinking, stimulating analytic insights.

The main categorizing strategy in qualitative research is coding. This is quite different from coding in quantitative research, which consists of applying a preestablished set of categories to the data according to explicit, unambiguous rules, with the primary goal being to generate frequency counts of the items in each category. In qualitative research, the goal of coding is not to count things, but to "fracture" (Strauss, 1987, p. 29) the data and rearrange them into categories that facilitate comparison between things in the same category and that aid in the development of theoretical concepts. Another form of categorizing analysis involves organizing the data into broader themes and issues.

An important set of distinctions in planning your categorizing analysis is among what I call "organizational," "substantive," and "theoretical" categories. Although these are not completely separate in practice, and intermediate categories are common, I think the conceptual distinction is valuable.

Organizational categories are broad areas or issues that you establish prior to your interviews or observations, or that could usually have been anticipated. McMillan and Schumacher (2001) referred to these as "topics" rather than categories, stating that "A topic is the descriptive name for the subject matter of the segment. You are not, at this time, asking 'What is said?' which identifies the meaning of the segment" (p. 469). In a study of elementary school principals' practices of retaining children in a grade, examples of such categories are "retention," "policy," "goals," "alternatives," "and "consequences" (p. 470). Organizational categories function primarily as "bins" for sorting the data for further analysis. They may be useful as chapter or section headings in presenting your results, but they don't help much with the actual work of making sense of what's going on (cf. Coffey & Atkinson, 1996, pp. 34–35).

This latter task requires substantive and/or theoretical categories, ones that provide some insight into what's going on. These latter categories can often be seen as subcategories of the organizational ones, but they are generally *not* subcategories that, in advance, you could have known would be significant, unless you are already fairly familiar with the kind of participants or setting you're studying or are using a well-developed theory. They implicitly make some sort of claim about the topic being studied—that is, they could be *wrong,* rather than simply being conceptual boxes for holding data.

Substantive categories are primarily *descriptive,* in a broad sense that includes description of participants' concepts and beliefs; they stay close to the data categorized, and don't inherently imply a more abstract theory. In the study of grade retention mentioned previously, examples of substantive categories would be "retention as failure," "retention as a last resort," "self-confidence as a goal," "parent's willingness to try alternatives," and "not being in control (of the decision)" (drawn from McMillan & Schumacher, 2001, p. 472). Categories taken from participants' own words and concepts (what are generally called "emic" categories) are usually substantive, but many substantive categories are not emic, being based on the *researcher's* description of what's going on. Substantive categories are often inductively developed through a close "open coding" of the data (Strauss & Corbin, 1998). They can be used in *developing* a more general theory of what's going on, but they don't *depend on* this theory.

Theoretical categories, in contrast, place the coded data into a more general or abstract framework. These categories may be derived either from prior theory or from an inductively developed theory (in which case the concepts and the theory are usually developed concurrently). They usually represent the *researcher's* concepts (what are called "etic" categories), rather than denoting participants' own concepts. For example, the categories "nativist," "remediationist," and "interactionist," used to classify teachers' beliefs about grade retention in terms of prior analytic dimensions (Smith & Shepard, 1988), would be theoretical.

The distinction between organizational categories and substantive or theoretical categories is important because some beginning qualitative researchers use mostly organizational categories to formally analyze their data, and don't systematically develop and apply substantive or theoretical categories in developing their conclusions. The more data you have, the more important it is to create the latter types of categories; with any significant amount of data, you can't hold all of the data relevant to particular substantive or theoretical points in your mind, and need a formal organization and retrieval system. In addition, creating substantive categories is particularly important for ideas (including participants' ideas) that don't fit into existing organizational or theoretical categories; such substantive ideas may get lost, or never developed, unless they can be captured in explicit categories. Consequently, you need to include in your design (and proposal) strategies for developing substantive and theoretical categories.

Connecting strategies operate quite differently from categorizing ones such as coding. Instead of "fracturing" the initial text into discrete segments and re-sorting it into categories, connecting analysis attempts to understand the data (usually, but not necessarily, an interview transcript or other textual material) in context, using various methods to identify the relationships among the different elements of the text (Atkinson, 1992; Coffey & Atkinson, 1996; Mishler, 1986). Examples of connecting strategies include some case studies (Stake, 1995), profiles and vignettes (Seidman, 1998), some types of discourse analysis (Gee, Michaels, & O'Connor, 1992) and narrative analysis (Coffey & Atkinson, 1996; Riessman, 1993), reading for "voice" (Brown, 1988), and ethnographic microanalysis (Erickson, 1992). What all of these strategies have in common is that they do not focus primarily on *similarities* that can be used to sort data into categories independently of context, but instead look for relationships that *connect* statements and events within a context into a coherent whole.

The identification of connections among different categories and themes can also be seen as a connecting step in analysis (Dey, 1993), but it is a broader one that works with the results of a prior categorizing analysis. This connecting step is necessary for building theory, a primary goal of analysis. However, it cannot recover the contextual ties that were lost in the original categorizing analysis. A purely connecting analysis, on the other hand, is limited to understanding particular individuals or situations, and cannot develop a more general theory of what's going on. The two strategies need each other to provide a well-rounded account (Maxwell & Miller, n.d.).

The difference between categorizing and connecting strategies has important consequences for your overall design. A research question that asks about the way events in a specific context are connected cannot be answered by an exclusively categorizing analytic strategy (see Example 5.2). Conversely, a question about similarities and differences across settings or individuals cannot be answered by an exclusively connecting strategy. Your analysis strategies have to be compatible with the questions you are asking.

EXAMPLE 5.2

A Mismatch Between Questions and Analysis

Mike Agar (1991) was once asked by a foundation to review a report on an interview study that it had commissioned, investigating how historians worked. The researchers had used the computer program The Ethnograph to segment and code the interviews by topic and collect together all the segments on the same topic; the report discussed each of these topics, and provided examples of how the historians talked about these. However, the foundation felt that the report hadn't really answered its questions, which had to do with how individual historians thought about their work—their theories about how the different topics were connected and the relationships they saw between their thinking, actions, and results.

Answering the latter question would have required an analysis that elucidated these connections in each historian's interview. However, the categorizing analysis on which the report was based fragmented these connections, destroying the contextual unity of each historian's views and allowing only a collective presentation of shared concerns. Agar argued that the fault was not with The Ethnograph, which is extremely useful for answering questions that require categorization, but with its misapplication. As he commented, "The Ethnograph represents a *part of* an ethnographic research process. When the part is taken for the whole, you get a pathological metonym that can lead you straight to the right answer to the wrong question" (p. 181).

What do I need to know?	*Why do I need to know this?*	*What kind of data will answer the questions?*	*Where can I find the data?*	*Whom do I contact for access?*	*Time lines for acquisition*
What are the truancy rates for American Indian students?	To assess the impact of attendance on American Indian students' persistence in school	Computerized student attendance records	Attendance offices; assistant principal's offices for all schools	Mr. Joe Smith, high school assistant principal; Dr. Amanda Jones, middle school principal	August: Establish student database October: Update June: Final tally
What is the academic achievement of the students in the study?	To assess the impact of academic performance on American Indian students' persistence in school	Norm- and criterion-referenced test scores; grades on teacher-made tests; grades on report cards; student portfolios	Counseling offices	High school and middle school counselors; classroom teachers	Compilation #1: End of semester Compilation #2: End of school year
What is the English-language proficiency of the students?	To assess the relationship between language proficiency, academic performance, and persistence in school	Language-assessment test scores; classroom teacher attitude surveys; ESL class grades	Counseling offices; ESL teachers' offices	Counselors' test records; classroom teachers	Collect test scores Sept. 15 Teacher survey, Oct. 10–15 ESL class grades, end of fall semester and end of school year
What do American Indian students dislike about school?	To discover what factors lead to antischool attitudes among American Indian students	Formal and informal student interviews; student survey	Homeroom classes; meetings with individual students	Principals of high school and middle schools; parents of students; homeroom teachers	Obtain student and parent consent forms, Aug.–Sept. Student interviews, Oct.–May 30 Student survey, first week in May

Figure 5.1 Data-Planning Matrix for a Study of American Indian At-Risk High School Students*

What do I need to know?	*Why do I need to know this?*	*What kind of data will answer the questions?*	*Where can I find the data?*	*Whom do I contact for access?*	*Time lines for acquisition*
What do students plan to do after high school?	To assess the degree to which coherent post–high school career planning affects high school completion	Student survey; follow-up survey of students attending college and getting jobs	Counseling offices; Tribal Social Services office; Dept. of Probation; Alumni Association	Homeroom teachers; school personnel; parents; former students; community social service workers	Student survey, first week in May Follow-up survey, summer and fall
What do teachers think about their students' capabilities?	To assess teacher expectations of student success	Teacher survey; teacher interviews	—	Building principals; individual classroom teachers	Teacher interviews, November (subgroup) Teacher survey, April (all teachers)
What do teachers know about the home culture of their students?	To assess teachers' cultural awareness	Teacher interviews; teacher survey; logs of participation in staff development activities	Individual teachers' classrooms and records	Building principals; individual classroom teachers; assistant superintendent for staff development	Teacher interviews, November (subgroup) Teacher survey, April (all teachers)
What do teachers do to integrate knowledge of the student's home culture community into their teaching?	To assess the degree of discontinuity between school culture and home culture	Teachers' lesson plans; classroom observations; logs of participation in staff development activities	Individual teachers' classrooms and records	Building principals; individual classroom teachers; assistant superintendent for staff development	Lesson plans, December–June Observations, Sept. 1–May 30 Staff development, June logs

SOURCE: Adapted from *Ethnography and Qualitative Design in Educational Research* (2nd ed.), by M. D. LeCompte and J. Preissle, 1993, San Diego: Academic Press.
NOTE: *Research problem: To what extent do various at-risk conditions contribute to dropping out for American Indian students?

LINKING METHODS AND QUESTIONS

To design a workable and productive study, and to communicate this design to others, you need to create a *coherent* design, one in which the different methods fit together compatibly, and in which they are integrated with the other components of your design. The most critical connection is with your research questions, but, as discussed previously, this is primarily an *empirical* connection, not a logical one. If your methods won't provide you with the data you need to answer your questions, you need to change either your questions or your methods.

A useful tool in assessing this compatibility is a matrix in which you list your questions and identify how each of the components of your methods will help you to get the data to answer these questions. Such a matrix displays the *logic* of your methods decisions. I have included one example of how such a matrix can be used[5] (Figure 5.1); such a matrix can be valuable as an appendix to a research proposal. Following this, I have provided an exercise for you to develop a matrix for your own study.

EXERCISE 5.2

Questions and Methods Matrix

This exercise has two purposes. The first is for you to link your research questions and your research methods—to display the logical connections between your research questions and your selection, data collection, and data analysis decisions. The second purpose is to gain experience in using matrices as a tool; matrices are useful not only for research design, but also for ongoing monitoring of selection and data collection (see Miles & Huberman, 1994, p. 94) and for data analysis.

Doing this exercise can't be a mechanical process; it requires thinking about *how* your methods can provide answers to your research questions. One way to do this is to start with your questions and ask what data you would need, how you could get these data, and how you could analyze them in order to answer these questions. You can also work in the other direction: Ask yourself *why* you want to collect and analyze the data in the way you propose—what will you learn from this? Then examine these connections between your research questions and your methods, and work on displaying these connections in a matrix. Doing this may require you to revise your questions, your planned methods, or both. Keep in mind that this exercise is intended to help you *make* your methods decisions, not as a final formulation of these.

The exercise has two parts:

1. Construct the matrix itself. Your matrix should include columns for research questions, selection decisions, data collection methods, and kinds of analyses, but you can add any other columns you think would be useful in explaining the logic of your design.
2. Write a brief narrative *justification* for the choices you make in the matrix. One way to do this is as a separate discussion, by question, of the rationale for your choices in each row; another way is to include this as a column in the matrix itself (as in Figure 5.1).

NOTES

1. This is simply another application of the variance versus process distinction discussed earlier. Rather than focusing on the *degree* of prestructuring and its consequences (treating prestructuring as a variable that can affect other variables), I am concerned with the *ways* that prestructuring is employed in actual studies and *how* it affects other aspects of the design.

2. However, he also dismissed as invalid a widely used argument for the generalizability of data from a convenience sample—a similarity between some demographic characteristics of the sample and of the population as a whole.

3. This process resembles that used for stratified random sampling; the main difference is that the final selection is purposeful rather than random.

4. Strauss (1987; Strauss & Corbin, 1990, pp. 176–193) developed a strategy that he called "theoretical sampling," which is a variation on this third approach. Theoretical sampling is driven by the theory that is inductively developed *during* the research (rather than by prior theory); it selects for examination those particular settings, individuals, events, or processes that are most relevant to the emerging theory.

5. There are numerous examples of other types of matrices in Miles and Huberman (1994).

6

Validity

How Might You Be Wrong?

In the movie *E. T. the Extra-Terrestrial,* there is a scene near the end of the film where the hero and his friends are trying to rescue ET and help him return to his spaceship. One of the boys asks, "Can't he just beam up?" The hero gives him a disgusted look and replies, "This is reality, Fred."

Validity, like getting to ET's spaceship, is the final component of your design. And as with ET's dilemma, there is no way to "beam up" to valid conclusions. This is reality. The validity of your results is not guaranteed by following some prescribed procedure. As Brinberg and McGrath (1985) put it, "Validity is not a commodity that can be purchased with techniques" (p. 13). Instead, it depends on the relationship of your conclusions to reality, and there are no methods that can completely assure that you have captured this.

The view that methods *could* guarantee validity was characteristic of early forms of positivism, which held that scientific knowledge could ultimately be reduced to a logical system that was securely grounded in irrefutable sense data. This position has been abandoned by philosophers, although it still informs many research methods texts (Phillips & Burbules, 2000, pp. 5–10). Validity is a goal rather than a product; it is never something that can be proven or taken for granted. Validity is also relative: It has to be assessed in relationship to the purposes and circumstances of the research, rather than being a context-independent property of methods or conclusions. Finally, validity threats are made implausible by *evidence,* not methods; methods are only a way of getting evidence that can help you rule out these threats.

The realist claim that validity can't be assimilated to methods is one of the two main reasons that, in the model presented here, I have made validity a distinct component of qualitative design, separate from methods. The second reason is pragmatic: that validity is generally acknowledged to be a key *issue* in research design, and I think it's important that it be explicitly addressed. Przeworski and Salomon (1988) identified, as one of the three questions that proposal readers seek answers to, "How will we know that the conclusions are valid?" And Bosk (1979) stated that "All fieldwork done by a single field-worker invites the question, Why should we believe it?" (p. 193). A lack of attention to validity threats is a common reason for the rejection of research proposals. Making validity an explicit component of design can help you to address this issue.

THE CONCEPT OF VALIDITY

In this book, I use validity in a fairly straightforward, commonsense way to refer to the correctness or credibility of a description, conclusion, explanation, interpretation, or other sort of account. I think that this commonsense use of the term is consistent with the way it is generally used by qualitative researchers, and does not pose any serious philosophical problems.[1] This use of the term "validity" does not imply the existence of any "objective truth" to which an account can be compared. However, the idea of "objective truth" isn't essential to a theory of validity that does what most researchers want it to do, which is to give them some grounds for distinguishing accounts that are credible from those that are not. Nor are you required to attain some ultimate truth in order for your study to be useful and believable.

Geertz (1973) told the story of a British gentleman in colonial India who, upon learning that the world rested on the backs of four elephants, who in turn stood on the back of a giant turtle, asked what the turtle stood on. Another turtle. And that turtle? "Ah, Sahib, after that it is turtles all the way down" (p. 29). Geertz's point is that there is no "bottom turtle" of ethnographic interpretation, that cultural analysis is essentially incomplete. While I accept Geertz's point, I would emphasize a different lesson: that you do not have to get to the bottom turtle to have a valid conclusion. You only have to get to a turtle you can stand on securely.

As Campbell (1988), Putnam (1990), and others have argued, we don't need an observer-independent "gold standard" to which we can compare our accounts to see if they are valid. All we require is the possibility of *testing* these accounts against the world, giving the phenomena that we are trying to understand the chance to prove us wrong. A key concept for validity is thus the validity *threat:* a way you might be wrong. These threats are often conceptualized as alternative explanations, or what Huck and Sandler (1979) called "rival hypotheses." Validity, as a component of your research design, consists of the strategies you use to identify and try to rule out these threats.

There are important differences between quantitative and qualitative designs in the ways they typically deal with validity threats. Quantitative and experimental researchers generally attempt to design, in advance, controls that will deal with both anticipated and unanticipated threats to validity. These include control groups, statistical control of extraneous variables, randomized sampling and assignment, the framing of explicit hypotheses in advance of collecting the data, and the use of tests of statistical significance. These prior controls deal with most validity threats in an anonymous, generic fashion; as Campbell put it, "randomization purports to control an infinite number of 'rival hypotheses' *without specifying what any of them are*" (1984, p. 8).

Qualitative researchers, on the other hand, rarely have the benefit of previously planned comparisons, sampling strategies, or statistical manipulations that "control for" plausible threats, and must try to rule out most validity threats after the research has begun, using evidence collected during the research itself to make these "alternative hypotheses" implausible. This strategy of addressing particular validity threats *after* a tentative account has been developed, rather than by attempting to eliminate such threats through prior features of the research design, is, in fact, more fundamental to the scientific method than is the latter approach (Campbell, 1988; Platt, 1964). However, this approach requires you to identify the *specific* threat in question and to develop ways to attempt to rule out that particular threat.

This conception of validity threats and how they can be dealt with is a key issue in a qualitative research proposal. Many qualitative proposal writers make the mistake of talking about validity only in general, theoretical terms, presenting abstract strategies such as "bracketing," "member checks," and "triangulation" that will supposedly protect their studies from invalidity. Such presentations often appear to be "boilerplate"—language that has been borrowed from methods books or successful proposals, without any demonstration that the author has thought through how these strategies will actually be applied in the proposed study. These sections of the proposal often remind me of magical charms that are intended to drive away evil; they lack any discussion of how these strategies will work in practice, and their use seems to be based largely on faith in their supernatural powers.

In contrast, the main emphasis of a qualitative proposal ought to be on how you will rule out *specific* plausible alternatives and threats to your interpretations and explanations. Citations of authorities and invocation of standard approaches are less important than providing a clear argument that the approaches described will adequately deal with the particular threats in question, in the context of the study being proposed. Martha Regan-Smith's proposal (see the Appendix) provides a good example of this.

TWO SPECIFIC VALIDITY THREATS: BIAS AND REACTIVITY

I argued previously that qualitative researchers generally deal with validity threats as particular events or processes that could lead to invalid conclusions, rather than as generic "variables" that need to be controlled. It clearly is impossible for me to list all, or even the most important, validity threats to the conclusions of a qualitative study, as Cook and Campbell (1979) attempted to do for quasi-experimental studies. What I want to do instead is to discuss two broad types of threats to validity that are often raised in relation to qualitative studies: researcher bias and the effect of the researcher on the individuals studied, often called reactivity.

Researcher "Bias"

Two important threats to the validity of qualitative conclusions are the selection of data that fit the researcher's existing theory or preconceptions and the selection of data that "stand out" to the researcher (Miles & Huberman, 1994, p. 263; Shweder, 1980). Both of these involve the subjectivity of the researcher, a term that most qualitative researchers prefer to "bias." As discussed in Chapters 2 and 3, it is impossible to deal with these issues by eliminating the researcher's theories, beliefs, and perceptual "lens." Qualitative research is not primarily concerned with eliminating *variance* between researchers in the values and expectations they bring to the study, but with understanding how a *particular* researcher's values and expectations influence the conduct and conclusions of the study (which may be either positive or negative) and avoiding the negative consequences. Explaining your possible biases and how you will deal with these is a key task of your research proposal. As one qualitative researcher, Fred Hess, phrased it, validity in qualitative research is not the result of indifference, but of integrity (personal communication).

Reactivity

The influence of the researcher on the setting or individuals studied, generally known as "reactivity," is a second problem that is often raised about qualitative studies. Trying to "control for" the effect of the researcher is appropriate to a quantitative, "variance theory" approach, in which the goal is to prevent researcher *variability* from being an unwanted cause of variability in the outcome variables. However, eliminating the *actual* influence of the researcher is impossible (Hammersley & Atkinson, 1995), and the goal in a qualitative study is not to eliminate this influence, but to understand it and to use it productively.

For participant observation studies, reactivity is generally *not* as serious a validity threat as some people believe. Becker (1970, pp. 45–48) pointed out that, in natural settings, an observer is generally much less of an influence on participants' behavior than is the setting itself (though there are clearly exceptions to this, such as situations in which illegal behavior occurs). For interviews, in contrast, reactivity—more correctly, what Hammersley and Atkinson (1995) called "reflexivity," the fact that the researcher is part of the world he or she studies—is a powerful and inescapable influence; what the informant says is *always* influenced by the interviewer and the interview situation. While there are some things you can do to prevent the more undesirable consequences of this (such as avoiding leading questions), trying to "minimize" your effect is not a meaningful goal for qualitative research. As discussed previously for "bias," what is important is to understand *how* you are influencing what the informant says, and how this affects the validity of the inferences you can draw from the interview.

VALIDITY TESTS: A CHECKLIST

Although methods and procedures do not guarantee validity, they are nonetheless essential to the process of ruling out validity threats and increasing the credibility of your conclusions. For this reason, I will provide a checklist of some of the most important strategies that can be used for this purpose. Miles and Huberman (1994, p. 262) included a more extensive list, having some overlap with mine, and other lists can be found in Becker (1970), Kidder (1981), Lincoln and Guba (1985), and Patton (1990). What follows is not a complete compilation of what these authors said—I urge you to consult their discussions—but simply what I see as most important (Maxwell, 2004c).

The overall point I want to make about these strategies is that they primarily operate not by *verifying* conclusions, but by *testing* the validity of your conclusions and the existence of potential threats to those conclusions (Campbell, 1988). The fundamental process in all of these tests is looking for evidence that could challenge your conclusions or make the potential threats implausible.

Keep in mind that these strategies work only if you actually *use* them. Putting them in your proposal as though they were magical spells that could drive away the validity threats (and criticism of the proposal) won't do the job; you will need to demonstrate that you have thought through how you can effectively use them in your own study. Not every strategy will work in a given study, and even trying to apply all the ones that are feasible might not be an efficient use of your time. As noted previously, you need to think in terms of *specific* validity threats, and what strategies are best able to deal with these.

1. Intensive, Long-Term Involvement

Becker and Geer (1957) claimed that long-term participant observation provides more complete data about specific situations and events than any other method. Not only does it provide more, and more different kinds, of data, but also the data are more direct and less dependent on inference. Repeated observations and interviews, as well as the sustained presence of the researcher in the setting studied, can help to rule out spurious associations and premature theories. They also allow a much greater opportunity to develop and test alternative hypotheses during the course of the research. For example, Becker (1970, pp. 49–51) argued that his lengthy participant observation research with medical students not only allowed him to get beyond their public expressions of cynicism about a medical career and uncover an idealistic perspective, but also enabled him to understand the processes by which these different views were expressed in different social situations, and how students dealt with the conflicts between these perspectives.

2. "Rich" Data

Both long-term involvement and intensive interviews enable you to collect "rich" data, data that are detailed and varied enough that they provide a full and revealing picture of what is going on[2] (Becker, 1970, pp. 51–62). In interview studies, such data generally require verbatim transcripts of the interviews, not just notes on what you felt was significant. For observation, rich data are the product of detailed, descriptive note taking (or videotaping and transcribing) of the specific, concrete events that you observe (Emerson, Fretz, & Shaw, 1995). Becker (1970) argued that such data

> counter the twin dangers of respondent duplicity and observer bias by making it difficult for respondents to produce data that uniformly support a mistaken conclusion, just as they make it difficult for the observer to restrict his observations so that he sees only what supports his prejudices and expectations. (p. 53)

Martha Regan-Smith's study of medical school teaching (see the Appendix) relied on lengthy observation and detailed field notes recording the teachers' actions in classes and students' reactions to these. In addition, she conducted and transcribed numerous interviews with students, in which they explained in detail not only what it was that the exemplary teachers did that increased their learning, but how and why these teaching methods were beneficial. This set of data provided a rich, detailed grounding for, and test of, her conclusions.

3. Respondent Validation

Respondent validation (Bryman, 1988, pp.78–80; Lincoln & Guba, 1985, referred to this as "member checks") is systematically soliciting feedback about your data and conclusions from the people you are studying. This is the single most important way of ruling out the possibility of misinterpreting the meaning of what participants say and do and the perspective they have on what is going on, as well as being an important way of identifying your own biases and misunderstandings of what you observed. However, participants' feedback is no more inherently valid than their interview responses; both should be taken simply as *evidence* regarding the validity of your account (cf. Hammersley & Atkinson, 1995). For a more detailed discussion of this strategy, see Bloor (1983), Bryman (1988, pp.78–80), Guba and Lincoln (1989), and Miles and Huberman (1994, pp. 242–243).

4. Intervention

Although some qualitative researchers see experimental manipulation as inconsistent with qualitative approaches (e.g., Lincoln & Guba, 1985), informal interventions are often used within traditional qualitative studies that lack a formal "treatment." For example, Goldenberg (1992), in a study of two students' reading progress and the effect that their teacher's expectations and behavior had on this progress, shared his interpretation of one student's failure to meet these expectations with the teacher. This resulted in a change in the teacher's behavior toward the student, and a subsequent improvement in the student's reading. The intervention with the teacher, and the resulting changes in her behavior and the student's progress, supported Goldenberg's claim that the teacher's behavior, rather than her expectations of the student, was the primary cause of the student's progress or lack thereof. In addition, Goldenberg provided an account of the *process* by which the change occurred, which corroborated the identification of the teacher's behavior as the cause of the improvement in a way that a simple correlation could never do.

Furthermore, in field research, the researcher's presence is *always* an intervention in some ways, as discussed in Chapter 5, and the effects of this presence can be used to develop or test ideas about the group or topic studied. For example, J. Briggs (1970), in her study of an Eskimo family, used a detailed analysis of how the family reacted to her often inappropriate behavior as an "adopted daughter" to develop her theories about the culture and dynamics of Eskimo social relations.

5. Searching for Discrepant Evidence and Negative Cases

Identifying and analyzing discrepant data and negative cases is a key part of the logic of validity testing in qualitative research. Instances that cannot be accounted for by a particular interpretation or explanation can point up important defects in that account. However, there are times when an apparently discrepant instance is not persuasive, as when the interpretation of the discrepant data is itself in doubt. Physics is full of examples of supposedly "disconfirming" experimental evidence that was later found to be flawed. The basic principle here is that you need to rigorously examine both the supporting and the discrepant data to assess whether it is more plausible to retain or modify the conclusion, being aware of all of the pressures to ignore data that do not fit your conclusions. Asking others for feedback is a valuable way to check your own biases and assumptions and flaws in your logic or methods. In particularly difficult cases, the best you may be able to do is to report the discrepant evidence and allow readers to evaluate this and draw their own conclusions (Wolcott, 1990).

6. Triangulation

Triangulation—collecting information from a diverse range of individuals and settings, using a variety of methods—was discussed in Chapter 5. This strategy reduces the risk of chance associations and of systematic biases due to a specific method, and allows a better assessment of the generality of the explanations that one develops. The most extensive discussion of triangulation as a validity-testing strategy in qualitative research is by Fielding and Fielding (1986).

One of Fielding and Fielding's key points is that it is not true that triangulation automatically increases validity. First, the methods that are triangulated may have the *same* biases and sources of invalidity, and thus provide only a false sense of security. For example, interviews, questionnaires, and documents are all vulnerable to self-report bias. Fielding and Fielding therefore emphasized the need to recognize the fallibility of *any* particular method or data and to triangulate in terms of *validity threats*. As argued previously, you should think about what particular sources of error or bias might exist, and look for specific ways to deal with this, rather than relying on your selection of methods to do this for you. In the final analysis, validity threats are ruled out by *evidence,* not methods.

7. Quasi-Statistics

Many of the conclusions of qualitative studies have an implicit quantitative component. Any claim that a particular phenomenon is typical, rare, or prevalent in the setting or population studied is an inherently quantitative claim, and requires some quantitative support. Becker (1970) coined the term "quasi-statistics" to refer to the use of simple numerical results that can be readily derived from the data. As he argued,

> One of the greatest faults in most observational case studies has been their failure to make explicit the quasi-statistical basis of their conclusions. (pp. 81–82)

Quasi-statistics not only allow you to test and support claims that are inherently quantitative, but also enable you to assess the *amount* of evidence in your data that bears on a particular conclusion or threat, such as how many discrepant instances exist and from how many different sources they were obtained. This strategy is used effectively in a classic participant-observation study of medical students by Becker, Geer, Hughes, and Strauss (1961), which presented more than 50 tables and graphs of the amount and distribution of observational and interview data supporting their conclusions.

8. Comparison

Explicit comparisons (such as between intervention and control groups) for the purpose of assessing validity threats are most common in quantitative, variance theory research, but there are numerous uses of comparison in qualitative studies, particularly in multicase or multisite studies. Miles and Huberman (1994, p. 254) provided a list of strategies for comparison, as well as advice on their use. Such comparisons (including comparisons of the same setting at different times) can address one of the main objections to using qualitative methods for understanding causality—their inability to explicitly address the "counterfactual" of what would have happened *without* the presence of the presumed cause (Shadish, Cook, & Campbell, 2002, p. 501).

In addition, single-setting qualitative studies, or interview studies of a relatively homogeneous group of interviewees, often incorporate less formal comparisons that contribute to the interpretability of the results. There may be a literature on "typical" settings or individuals of the type studied that makes it easier to identify the relevant characteristics and processes in an exceptional case and to understand their significance. In other instances, the participants in the setting studied may themselves have experience with other settings or with the same setting at an earlier time, and the researcher may be able to draw on this experience to identify the crucial factors and the effect that these have.

For example, Martha Regan-Smith's study of how exceptional medical school teachers help students to learn (see the Appendix) included only faculty who had won the "Best Teacher" award. From the point of view of quantitative design, this was an "uncontrolled" study, vulnerable to all of the validity threats identified by Campbell and Stanley (1963). However, both of the previously mentioned forms of implicit comparison were employed in the research. First, there is a great deal of published information about medical school teaching, and Regan-Smith was able to use both this background and her own extensive knowledge of medical schools to identify what it was that the teachers she studied did in their classes that was distinctive. Second, the students she interviewed explicitly contrasted these teachers with others whose classes they felt were not as helpful. In addition to these comparisons, the validity of her research conclusions depended substantially on a process approach; the students explained in detail not only *what* it was that the exemplary teachers did that increased their learning, but also *how* and *why* these teaching methods were beneficial. Many of these explanations were corroborated by Regan-Smith's own experiences as a participant-observer in these teachers' classes and by the teachers' explanations of why they taught the way they did.

EXERCISE 6.1

Identifying and Dealing With Validity Threats

1. What are the most serious validity threats (alternative explanations) that you need to be concerned with in your study? In other words, what are the main ways in which you might be mistaken about what's going on? Be as specific as you can, rather than just giving general categories. Also, think about *why* you believe these might be serious threats.

2. What could you do in your research design (including data collection and data analysis) to deal with these threats and increase the credibility of your conclusions? Start by brainstorming possible solutions, and then consider which of these strategies are *practical* for your study, as well as theoretically relevant.

Remember that some validity threats are unavoidable; you will need to acknowledge these in your proposal or in the conclusions to your study, but no one expects you to have airtight answers to *every* possible threat. The key issue is how plausible and how serious these unavoidable threats are.

GENERALIZATION IN QUALITATIVE RESEARCH

I have deliberately left generalization until the end, because I consider it a separate issue from validity proper. Qualitative researchers usually study a single setting or a small number of individuals or sites, using theoretical or purposeful rather than probability sampling, and they rarely make explicit claims about the generalizability of their accounts. However, it is important to distinguish between what I call "internal" and "external" generalizability (Maxwell, 1992). Internal generalizability refers to the generalizability of a conclusion *within* the setting or group studied, while external generalizability refers to its generalizability beyond that setting or group. Internal generalizability is clearly a key issue for qualitative case studies; it corresponds to what Cook and Campbell (1979) called "statistical conclusion validity" in quantitative research. The descriptive, interpretive, and theoretical validity of the conclusions of a case study all depend on their internal generalizability to the case as a whole. If you are studying the patterns of interaction between the teacher and students in a single classroom, your account of that classroom as a whole is seriously jeopardized if you have selectively focused on particular students or kinds of interactions and ignored others.

In contrast, external generalizability is often not a crucial issue for qualitative studies. Indeed, the value of a qualitative study may depend on its *lack* of external generalizability in the sense of being representative of a larger population, as discussed in Chapter 5; it may provide an account of a setting or population that is illuminating as an extreme case or "ideal type." Eliot Freidson, in his study of social control among physicians (Example 3.2), selected an atypical group practice, one that was staffed by physicians who were better trained and whose views were more progressive than usual and that was structured precisely to deal with the issues he was addressing. He argued that his study made an important contribution to theory and policy precisely because this was a group for whom social controls on practice should have been most likely to be effective. The failure of such controls in this case not only highlights a social process that is likely to exist in other groups, but also provides a more persuasive argument for the unworkability of such controls than would a study of a "representative" group.

This does not mean that qualitative studies are never generalizable beyond the setting or informants studied. First, qualitative studies often have what Judith Singer (personal communication) called "face generalizability"; there is no obvious reason *not* to believe that the results apply more generally. Second, the generalizability of qualitative studies is usually based not on explicit sampling of some defined population to which the results can be extended, but on the development of a *theory* that can be extended to other cases (Becker, 1991; Ragin, 1987; Yin, 1994). Third, Hammersley (1992, pp. 189–191) and Weiss (1994, pp. 26–29) listed a number of features that lend plausibility to generalizations from case studies or nonrandom samples, including respondents' own assessments of generalizability, the similarity of dynamics and constraints to other situations, the presumed depth or universality of the phenomenon studied, and corroboration from other studies. All of these characteristics can provide credibility to generalizations from qualitative studies, but none permits the kinds of precise extrapolation of results to defined populations that probability sampling allows.

NOTES

1. I present the philosophical argument that informs these statements elsewhere (Maxwell, 1992, 2002, 2004c). I also think that the concept of validity presented here is compatible with some "postmodern" approaches to validity (e.g., Kvale, 1989; Lather, 1993; cf. Maxwell, 1995, 2004b).

2. Some qualitative researchers refer to these sorts of data as "thick description," a phrase coined by the philosopher Gilbert Ryle (1949) and applied to ethnographic research by Geertz (1973). This is not what either Ryle or Geertz meant by the phrase. "Thick description," as Geertz used it, is description that incorporates the intentions of the actors and the codes of signification that give their actions meaning for them, what anthropologists call an *emic* account. It has nothing to do with the amount of detail provided. (For a more detailed discussion of this issue, see Maxwell, 1992, pp. 288–289.)

7

Research Proposals

Presenting and Justifying a Qualitative Study

Catherine the Great of Russia once decided to take a cruise down the Danube to view that part of her empire. Her prime minister, Grigory Potemkin, knowing that the poverty of the region would not be pleasing to the empress, allegedly built fake villages along the banks of the river and forcibly staffed these with cheering peasants, in order to impress the empress with how prosperous and thriving the area was. The term "Potemkin village" has since come to be used to refer to "an impressive facade or show designed to hide an undesirable fact or condition" (*Webster's New Collegiate Dictionary*).

Some proposals are, to a significant extent, Potemkin villages. What is presented does not reflect what the author actually believes or plans to do, but is fabricated in order to get approval or money for the study or to fit what the writer thinks a proposal ought to say. Such proposals are frequently the result of the writer not having worked out (or worse, not having understood the *need* to work out) the actual design of the study, and thus having to substitute a fake design for this. Aside from the fact that reviewers are usually fairly good at detecting such facades, the most serious danger of a "Potemkin village" proposal is that you may be taken in by your own fabrication, thinking that you have, in fact, solved your design problems, and thus ignoring your actual theories, goals, questions, and situation and the implications of these—your real research design. An ignorance of, or refusal to acknowledge, this real design and the conditions that affect it are certain to cause problems when you actually try to do the study.

Of course, as discussed in previous chapters, your research design will evolve as you conduct the study, and therefore a proposal for a qualitative study can't present an exact specification of what you will do. However, this is no excuse for not developing the design for your study in as much detail as you can at this point, or for failing to clearly communicate this design. In your proposal, you simply need to explain the kinds of flexibility that your study requires, and indicate, as best you can, *how* you will go about making future design decisions. For dissertation proposals, your committee often wants to see that you have demonstrated the *ability* to design a coherent and feasible study, providing evidence that you are aware of the key issues in your proposed research and ways of dealing with these, rather than requiring a completely worked-out design.

In this chapter, I want to explain the connections between a study's research design and an effective proposal for that study, and to provide some guidelines and advice on how to accomplish the transition from design to proposal. I believe that the model that I have presented in this book simplifies and facilitates this transition, and provides a useful framework for thinking about proposal structure and content. Much more detailed and specific advice on proposal writing is provided by Locke, Spirduso, and Silverman (2000).

I will begin with the purposes and structure of a research proposal, and then take up the ways in which the design of your study connects to these purposes and structure. Finally, I will discuss the specific parts of a proposal and the key issues that a proposal for qualitative research needs to address.

THE PURPOSE OF A PROPOSAL

The structure of a proposal isn't an arbitrary set of rules; it's closely tied to the *purpose* of a proposal. This purpose is so fundamental that when you are working on a proposal, you should post it above your desk or computer: *The purpose of a proposal is to explain and justify your proposed study to an audience of nonexperts on your topic.*

There are four key concepts in this statement:

1. *Explain.* You want your readers to clearly understand what you plan to do. Locke et al. (2000) emphasized that "advisors and reviewers misunderstand student proposals far more often than they disagree with what is proposed" (p. 123). This observation is abundantly supported by my own experience, both with advising and reviewing student proposals and with submitting and reviewing grant proposals. In writing and editing your proposal, *clarity* is a primary goal.

2. *Justify.* You want the readers of your proposal to understand not only *what* you plan to do, but *why*—your rationale for how you plan to conduct the study. Proposals are often not accepted, even when the study is clearly described, because it isn't clear *why* the author wants to do the study a certain way. Your readers may not understand how your proposed methods will provide valid answers to your research questions, or how the questions address important issues or purposes. They may also question whether *you* have a good

reason for doing the study this way, or if you are simply using "boilerplate" language that you've borrowed from other studies.

3. *Your proposed study.* Your proposal should be about *your study,* not the literature, your research topic, or research methods in general. You should ruthlessly edit out anything in the proposal that does not directly contribute to the explanation and justification of your study. A proposal is no place to display your general knowledge of the literature on your topic, your theoretical or methodological sophistication, or your political views on the issues you plan to investigate;[1] this will generally annoy your reviewers, who are trying to determine if your proposed study makes sense.

Students sometimes focus their proposal on their planned *dissertation,* rather than on the research that they propose to do. They provide lengthy, chapter-by-chapter descriptions of what the dissertation will cover, and use language such as, "In my dissertation, I will discuss. . . ." While it can occasionally be helpful, in explaining and justifying your study, to refer to how you intend to present this in the dissertation, far more often these references to your dissertation are red herrings, interfering with your presentation of the actual research and its design.

4. *Nonexperts.* You can't assume any particular specialized knowledge on the part of your readers. Grant proposals in the social sciences and related fields are generally *not* assigned to readers on the basis of their expertise on your specific topic, and students often will have faculty reviewing their proposals who are not knowledgeable about the specific area of the proposed study. You need to carefully examine your proposal to make sure that everything in it will be clear to a nonspecialist. (The best way to do this is generally to *give* the proposal to some nonspecialists and ask them to tell you what isn't clear.)

THE PROPOSAL AS AN ARGUMENT

Another way of putting the points made previously is that a proposal is an argument *for* your study. It needs to explain the logic behind the proposed research, rather than simply describe or summarize the study, and to do so in a way that nonspecialists will understand. (It should *not,* however, attempt to defend your anticipated *conclusions;* doing so is almost certain to raise serious questions about your own biases.) Each piece of your proposal should form a clear part of this argument.

The essential feature of a good argument is *coherence,* and a proposal needs to be coherent in two different senses of this term. First, it has to *cohere*—flow logically from one point to the next, and hang together as an integrated whole.

The connections among different components of your design are crucial to this coherence. You need to understand *why* you're doing what you're doing, rather than blindly following rules, models, or standard practice. Examples 7.1 and 7.2 and Exercise 7.1 can help you to achieve this.

Second, it has to *be coherent*—to make sense to the reviewers. You need to put yourself in your readers' shoes, and think about how what you say will be understood by them. This requires avoiding jargon, unnecessarily complex style, and what Becker (1986) called "classy writing." A failure to achieve these two aspects of coherence is the source of the most common problems with proposals: Either they have inconsistencies or gaps in their reasoning, or they don't adequately communicate to the reviewers what the author wants to do and why, or both. Martha Regan-Smith's proposal (see the Appendix) is a good example of clear, straightforward language that avoids these problems.

THE RELATIONSHIP BETWEEN RESEARCH DESIGN AND PROPOSAL ARGUMENT

There are a number of questions that reviewers will be asking in reading your proposal, questions that the argument of your proposal needs to address. According to Locke et al. (2000),

> The author must answer three questions:
>
> 1. What do we already know or do? . . .
> 2. How does this particular question relate to what we already know or do? . . .
> 3. Why select this particular method of investigation? (pp. 17–18)

These questions emphasize the connections along one axis of my model of research design: the one that runs from the upper right to lower left of the diagram, consisting of the Conceptual Framework, Research Questions, and Methods (see Figure 7.1).

In contrast, Przeworski and Salomon (1988), in their suggestions for applicants seeking funding from the Social Science Research Council, stated that

> Every proposal reader constantly scans for clear answers to three questions:
>
> - What are we going to learn as the result of the proposed project that we do not already know?
> - Why is it worth knowing?
> - How will we know that the conclusions are valid? (p. 2)

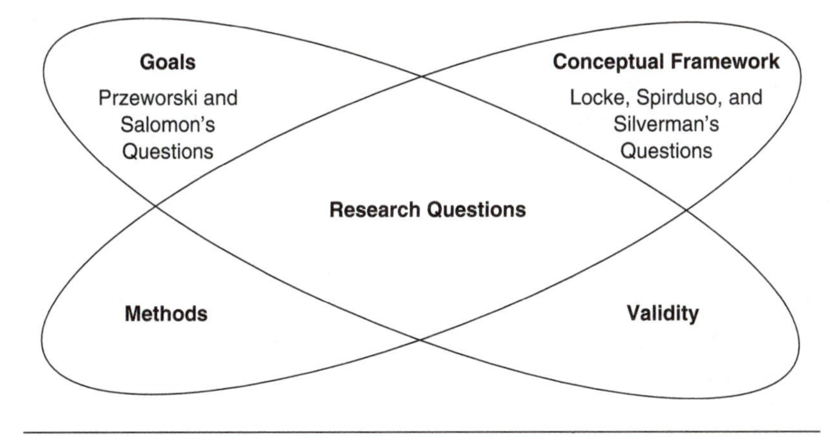

Figure 7.1 Relationship Between Research Design and Questions About Research Proposals

These questions, in contrast to those of Locke et al., emphasize the connections along the other axis of the model, the one that runs from the upper left to lower right of the diagram, including Goals, Research Questions, and Validity.

Thus, the relationships among the components of your research design constitute a crucial part of the argument of your proposal. These relationships provide the coherence that your argument depends on. Above all else, your proposal must convey to the readers what these connections are.

A MODEL FOR PROPOSAL STRUCTURE

The model of research design that I have presented in this book can be directly mapped onto one way of organizing a qualitative proposal. This format is not the only way to structure a proposal, but it is a fairly standard and generally understood format, and one that lends itself particularly well to communicating the design of a qualitative study. However, every university and funding source has its own requirements and preferences regarding proposal structure, and these must take precedence for your official proposal if they conflict with what I present here. I still recommend, though, that you use the structure I describe here as a first step in writing the proposal, even if you will eventually convert it into a different format. I have seen too many students become lost by trying to use a traditional proposal structure to *develop* their design, producing a repetitive, incoherent argument that fails to convey the real strengths of their research.

I will first display the relationships between research design and proposal structure in diagram form (Figure 7.2), and then go through each part of the proposal structure in detail, explaining how it relates to my model of research design. This explanation will make more sense if it is read in conjunction with Martha Regan-Smith's proposal for a qualitative study of what four outstanding medical school teachers do to help students learn basic science, and my commentary on this proposal (see the Appendix). What is important about the structure I describe isn't having separate sections with these names; this is simply a useful organizational tool that can be modified if it conflicts with the structure you are required to follow. The point is to organize the issues in a way that clearly communicates your research design and its justification.

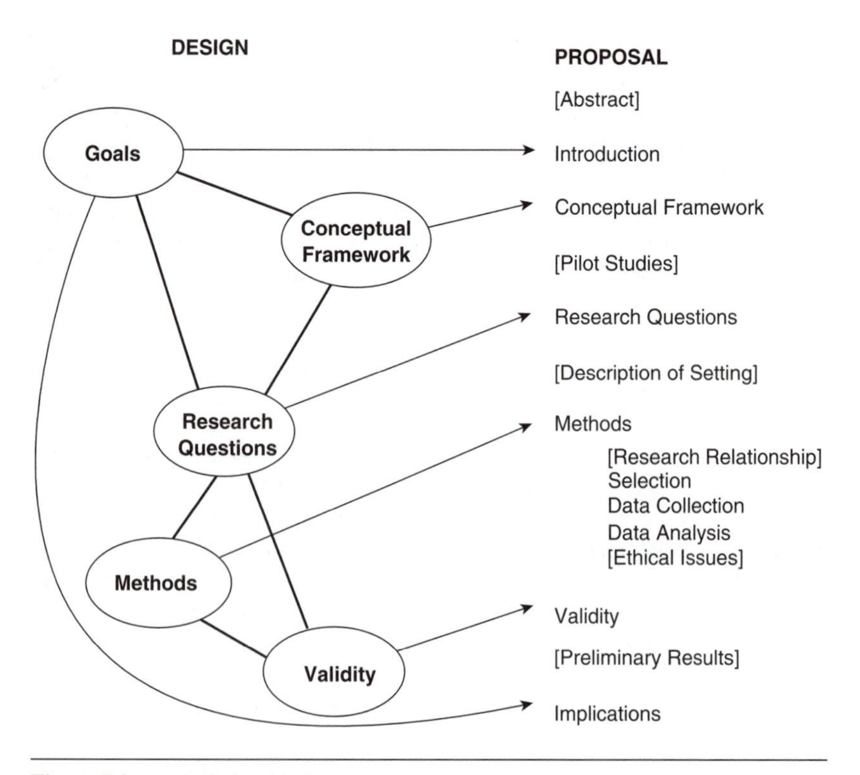

Figure 7.2 Relationship Between Research Design and Proposal Structure

1. Abstract

Not all proposals will require an abstract, but if you need to have one, this is the place to provide an overview and "road map," not just of the study itself, but also of the *argument* of your proposal. Your abstract should present in abbreviated form the actual argument for your research, not simply provide "placeholders" that will later be filled in with real content (Becker, 1986, pp. 50–53). Regan-Smith's abstract (see the Appendix) is a model for how to do this.

2. Introduction

The introduction to your proposal "sets the stage for your research, explaining . . . what you want to do and why" (Peters, 1992, p. 202). It should clearly present the goals of your study and the problem(s) it addresses, and give an overview of your main research questions and of the kind of study you are proposing. (A full presentation of your research questions is often better reserved until after the conceptual framework section, when the theoretical grounding of the questions will be clearer, but this is not an absolute rule.) It should also explain the structure of the proposal itself, if this could be confusing.

3. Conceptual Framework

This section is often called the "literature review"; this term is misleading, for reasons that I explained in Chapter 3, but you may need to use it, depending on whom the proposal is written for. This section of the proposal has two key functions. First, it needs to show how your proposed research fits into what is already known (its relationship to existing theory and research) and how it makes a contribution to our understanding of your topic (its intellectual goals). Second, it needs to explain the theoretical framework that informs your study. These functions are usually accomplished by discussing prior theory and research, but the point is not to *summarize* what's already been done in this field. Instead, it is to *ground* your proposed study in the relevant previous work, and to give the reader a clear sense of your theoretical approach to the phenomena that you propose to study. While some books on proposal writing argue that your proposal needs to demonstrate your familiarity with the literature in your field, I think this depends on your audience.

Insofar as your personal experience and knowledge form an important part of your conceptual framework, these should be discussed here; Martha Regan-Smith devotes a separate part of her context section to these. The key issue, again, is *relevance;* the connection of the experience and views discussed in this section to your study must be clear.[2]

Any pilot studies that you have done also need to be discussed in the proposal, explaining their implications for your research. This can be done in any of three places: at the end of the conceptual framework section; in a separate section immediately following the conceptual framework section; or, in some cases, after the presentation of your research questions, if a detailed grasp of these questions is important to understanding the pilot studies. Unless an important purpose of the pilot study was to try out the methods that you plan to use in your research, you should focus your discussion of your pilot studies on what you *learned* from them, rather than on the details of what you did.

4. Research Questions

As in my model of research design, the statement of your research questions is central to your proposal. Although you will usually present a brief statement of your main research questions in the introduction, I recommend putting a detailed discussion and explanation of these *after* the conceptual framework section. This is because the reasons for focusing on these particular questions may not be clear until the context of prior research, theory, and experience has been described. While you can create a short section just for your research questions, as Martha Regan-Smith did, you can also put them at the end of the conceptual framework section or at the beginning of the methods section.

The research questions section, in addition to stating your questions, should clarify two key points, if the answers to these are not obvious:

1. How your questions relate to prior research and theory, to your own experience and exploratory research, and to your goals.
2. How these questions form a coherent whole, rather than being a random collection of queries about your topic. Generally, a small number of clearly focused questions is far better than a larger number of questions that attempt to "cover the waterfront" on your topic. If you have more than two or three major questions, you need to think about whether some of these are best seen as subquestions of a broader question, or if your study is, in fact, attempting to do too much.

5. Research Methods[3]

Your proposal probably doesn't need to justify qualitative methods in general, unless you have reasons to think that this could be a concern for some readers.[4] You *do* need to explain and justify the *particular* methodological decisions you've made; for every decision, it should be clear *why* this is a reasonable choice. If you can't specify certain parts of your methods in advance (e.g., how many interviews you'll do), explain the *basis* on which you'll make your decision.

A description of the setting or social context of your study can be helpful in clarifying and justifying your choice of questions and methods. This description can be placed at the beginning of the methods section, or it can be a separate section just before or after the research questions. A proposal for funding will also need to explain what resources you already have and what ones you are requesting money for, your qualifications and experience, and your timetable and budget; some of this can be included in the methods section, but you will probably need additional sections as well.[5]

The methods section normally has several parts:

a. Research design in the typological sense. What kind of a study is this? This is not always necessary in a qualitative study, but it can sometimes be helpful to describe and justify the overall approach taken, for example, to explain why you have chosen to conduct a case study, or a comparison of two settings. If this doesn't require a detailed explanation, it can often be addressed in the introduction; if your research questions are closely tied to the kind of study you are doing (e.g., if you are comparing two settings and your questions focus on this comparison), this may be best addressed in the section on research questions.

b. The research relationship you establish with those you are studying. This is an important part of your design, as argued previously, but it is not usually an explicit part of a proposal. My advice is to discuss this relationship, particularly if it is an important and nonobvious source of information or insights, or if it raises potential data collection difficulties, ethical problems, or validity threats for the study.

c. Site and participant selection. It is important not simply to describe these, but also to explain *why* you have decided to study these particular settings or to interview this particular selection or number of people.

d. Data collection. How you will get the information you need to answer your research questions. This should include a description of the kinds of interviews, observations, or other methods you plan to use, how you will conduct these, and *why* you have chosen these methods. For both selection and data collection, practical considerations are often important, and your proposal should be candid about these, rather than ignoring them or concocting bogus theoretical justifications for decisions that are, in fact, practically based. If any of your decisions *are* based mainly on practical considerations (such as studying an institution where you have contacts and easy access), you need to deal at some point with any potential validity threats or ethical risks that this raises.

e. Data analysis. What you will do to make sense of the data you collect. Be explicit about how your data will be analyzed; specific examples are generally more useful than abstract descriptions. Also, be clear about how these analyses will enable you to answer your research questions; you may want to include a version of your questions and methods matrix (Example 5.1) to illustrate this.

Issues of ethics can be dealt with as part of the methods section, but if there are significant ethical questions that could be raised about your study, it may be better to have a separate ethics section, as Martha Regan-Smith does.

6. Validity

Validity issues are often dealt with under methods, but I recommend a separate validity section, for two reasons. The first is clarity—you can explain in one place how you will use different methods to address a single validity threat (a strategy discussed previously, known as triangulation), or how a particular validity issue will be dealt with through selection, data collection, and analysis decisions. The second reason is strategic: Devoting a separate section to validity emphasizes that you're taking validity seriously. For this and other issues in a proposal, it is often more important that your reviewers realize that you are *aware* of a particular problem, and are thinking about how to deal with it, than that you have an airtight plan for solving the problem.

A crucial issue in addressing validity is demonstrating that you will allow for the examination of competing explanations and discrepant data—that your research is not simply a self-fulfilling prophecy. Locke et al. (2000, pp. 87–89) provided a cogent discussion of "the scientific state of mind," and of the importance of developing alternative explanations and testing your conclusions. In my view, this issue is just as important for qualitative proposals as for quantitative ones.

7. Preliminary Results

If you have already begun your study, this is where you can discuss what you have learned so far about the practicality of your methods or tentative answers to your research questions. This discussion is often valuable in justifying the feasibility of your study and clarifying your methods, particularly your data analysis strategies; see Regan-Smith's proposal for an example of this.

8. Conclusion

This is where you pull together what you've said in the previous sections, remind your readers of the goals of the study and what it will contribute, and discuss its potential relevance and implications for the broader field(s) that it is situated in. This section should answer any "so what?" questions that might arise in reading the proposal. It is normally fairly short, a page or two at most.

9. References

This section should normally be limited to the references actually cited; unless you are directed otherwise, it should *not* be a bibliography of relevant literature.

10. Appendixes

These may include any of the following:

- a timetable for the research
- letters of introduction or permission
- questionnaires, interview guides, or other instruments
- a list of possible interviewees
- a schedule of observations
- descriptions of analysis techniques or software
- a matrix of relationships among questions, methods, data, and analysis strategies (see Figure 5.1)
- examples of observation notes or interview transcripts from pilot studies or completed parts of the study

The appendixes can also contain detailed explanations of things (e.g., a particular data collection or analysis technique, or background information about your informants or setting) that would require too much space to include in the body of the proposal.

The structure that I present here was originally developed for proposals of about 5,000 words (roughly 20 double-spaced pages). Different universities and funding sources have differing length requirements, some shorter and some longer than this. However, even if your submitted proposal needs to be shorter than this, I still recommend writing an initial draft of about 20 pages, because this is a good test of how well you have worked out your design. One student, whose 10-page proposal was approved by his committee, later said,

> I think it would have been better if I had done a more complete proposal. Even though I wasn't sure what form my research was going to take, I still should have

> spent more time planning. Then I would have had a greater feeling of confidence that I knew where I was going. (Peters, 1992, p. 201)

Once you are confident of your design and how to present this, you can edit this draft down to the required length. On the other hand, if you need to write a longer proposal than this, I advise *starting* with a draft of about this length, to help you develop your argument.

I want to emphasize that your research design can't be mechanically converted into a proposal. Your proposal is a document to *communicate* your design to someone else, and requires careful thinking, separate from the task of designing the research itself, about how best to accomplish this communication. To do this, you need take into account the particular audience for whom you are writing. Different universities, review boards, government agencies, and foundations all have their own perspectives and standards, and your design needs to be translated into the language and format required or expected by the people who will be reviewing the proposal. The structure I've presented here will usually be a good first approximation of what you need, but it may still require more or less adjustment to meet the expectations of your reviewers.

A useful step in moving from the generic proposal model presented here to a detailed proposal for your specific study is to prepare an outline of the *argument* of your proposal, in order to develop the sequence of points that you need to make to explain and justify your study. (Exercise 7.1 is an exercise in doing this.) This allows you to work specifically on the *logic* of the proposal, free from the constraints of style and grammatical structure. (For more on how to do this, see Becker, 1986, chap. 3.) As with concept maps, you can use this exercise in either of two ways—working to develop the logic from scratch and then converting this into a proposal, or taking a draft of your proposal, analyzing this to abstract the argument, and using this argument to revise the proposal. I've provided two examples of such outlines. Example 7.1 is my own outline of the argument of Martha Regan-Smith's proposal; it's fairly brief, but illustrates the basic idea. Example 7.2 is an outline that was actually written for a student's dissertation defense, but it's a good model (though a considerably more detailed one) of what I'm recommending.

As with my generic model for a proposal structure, I caution you not to use these outlines as *templates* for your own argument. Every study needs a different argument in order to adequately justify the research, and in developing this argument, you will need to work primarily from your *own* thinking about your study, not borrow someone else's. In particular, as I discuss in more detail in my comments on this, Martha Regan-Smith's study is investigating a topic on which little prior work has been done; your argument (and proposal) will almost certainly need to say more about existing theory and research.

EXAMPLE 7.1

The Argument of a Dissertation Proposal

Following is an outline of the argument of Martha Regan-Smith's proposal, which is presented in full in the Appendix. I have developed this outline from the proposal itself, so it's not a good example of the tentativeness that your own outline will probably display initially, but my main purpose here is to illustrate one way to outline your argument. Some of the points in this outline are implicit in the proposal, rather than explicit; the extent to which parts of your argument need to be explicitly stated in your proposal depends on what you can assume that your reviewers will easily infer or take for granted. Similarly, the outline itself is only a sketch of what would be necessary to completely justify the study; even in a full proposal, you will not be able to address every possible question about your research, and will have to focus on those issues that you think are most important for your audience.

Argument for a Study of How Basic Science Teachers Help Medical Students Learn

1. We need to better understand how basic science teachers in medical school help students learn.
   a. There has been an explosion in the amount of information that needs to be transmitted, with no increase in the time available to teach this.
   b. Medical students' performance on the basic science parts of licensing exams has declined.
   c. These facts have led to student disillusionment and cynicism, and to faculty concern.
2. We know little about how basic science teachers help students learn.
   a. Studies of science teachers in other settings don't necessarily apply to medical schools.
   b. Most research on basic science teaching has been quantitative, and doesn't elucidate how such teaching helps students learn.
   c. No one has asked medical students what teachers do that helps them to learn.
   d. The research I've already done indicates that students can identify what teachers do that helps them learn.

   e. Thus, a qualitative study of basic science teaching, focusing on student perspectives, can make an important contribution.

3. For these reasons, I propose to study four exemplary basic science teachers to understand:
   a. What they do that helps students to learn,
   b. How and why this is effective,
   c. What motivates these teachers, and
   d. The relationship between the students' and teachers' perspectives.
4. The setting and teachers selected are appropriate for this study.
   a. The medical school to be studied is typical, and my relationship with the school, teachers, and students will facilitate the study.
   b. The teachers selected are appropriate and diverse, and adding additional teachers would not contribute anything significant.
5. The methods I plan to use (participant observation and videotaping of lectures, student and teacher interviews, and documents) will provide the data I need to answer the research questions.
   a. Videotaping provides rich data on what happens in classes, and will be used to elicit reflection from the teachers.
   b. Interviews will be open-ended, and will incorporate questions based on the observations.
   c. The selection of students is guided by theoretical sampling, rather than statistical representativeness, in order to best understand how the teacher helps students.
6. Analysis will generate answers to these questions.
   a. My analysis will be ongoing and inductive in order to identify emergent themes, patterns, and questions.
   b. I will use coding and matrices for comparison across interviews, and interview summaries to retain the context of the data.
7. The findings will be validated by:
   a. Triangulating methods,
   b. Checking for alternative explanations and negative evidence,
   c. Discussing findings with teachers, students, and colleagues, and
   d. Comparing findings with existing theory.
   e. These methods, and others described earlier, will enable me to deal with the major validity threats to my conclusions: bias in the selection of teachers and students, and self-report bias for both.

8. The study poses no serious ethical problems.
   a. Teachers and students will be anonymous.
   b. I have taken measures to minimize the possible effect of my own authority.
9. Preliminary results support the practicability and value of the study.

EXAMPLE 7.2

An Outline of a Dissertation Proposal Argument

Using Computer-Mediated Communication Technology to Facilitate Students' Intellectual and Social Interaction and Learning in a Service-Learning Course

Mohamed Al-Ansari

I. Research Goal: This study attempts to accomplish the following goals

At the Theoretical Level:

- To verify the findings of previous research on the significance of computer-mediated communication (CMC) as a tool for fostering social and intellectual interaction.
- To bridge a gap in knowledge about the impact of CMC on students' learning in service-learning course context.

At the Practice Level:

- To enhance CMC use in this service-learning course.
  - To find out how the students in this service-learning course engaged in WebCT, and how it worked for them as means for social and intellectual interaction to promote peer support and learning.

At the Personal Level:

- This study meets a personal interest in studying the effectiveness of CMC across disciplines settings.
- A stepping stone toward further studies in CMC and service-learning.

II. Conceptual Framework: This study is informed by three fields of knowledge

Instructional Theory:

- Learning is a knowledge construction process.
- Learning cannot be understood independent of its social context.
- Interaction is two-way communication.
- Students' interaction is a key factor in students' learning.
  - Intellectual interaction is means for knowledge construction and validation.
  - Social interaction is means for collegial relationship building and peer support and reassurance.
- The role of the instructor changes from lecturer to facilitator of interaction between students for the purpose of knowledge construction and support.

CMC Research:

- Research on CMC suggests that with capability to hold textual discussion across time and space and to preserve discussion content, CMC was found capable to:
  - Foster intellectual interaction for knowledge construction and validation.
  - Foster reflective participation and discussion.
  - Foster social interaction for collegial relationship building and peer support.
- CMC advantages and limitations.
- Knowledge gap: Research on CMC is still at progress especially in domains where its effectiveness is not adequately investigated; such as in service-learning.

Service-Learning Research:

- Service-learning is field-based, inductive, and reflective learning.
- There is emphasis on students' need for interaction with peers and instructor as means for knowledge construction and socio-emotional support.
- The nature and structure of service-learning impose logistical challenges on packaging a course that satisfies both the need to allocate time for field-based learning as well as the need for classroom interaction.

My Beliefs:

- With the capability to foster social and intellectual interaction, CMC could be a "remedy" for the previously mentioned logistical challenge in service-learning courses.

- In this service-learning course WebCT (CMC) connected the students across time and space with more than 300 postings.
    - The quantity of postings doesn't give an accurate picture of how students interacted in WebCT.
    - It also doesn't tell how it worked for students as a method of knowledge building and validation as well as a means of peer support.
    - It's not clear how WebCT interaction impacted students' learning.

III. Research Questions

1. How did the students in this service-learning course perceive the intellectual and social interaction in WebCT (a) as means for providing them with support in regard to constructing and validating knowledge of service-learning, and (b) as means for providing them with socio-emotional support to cope with involvement in this course?
2. How did service-learning students engage in WebCT?
    a. What is the level of participation as measured by the frequency of length of postings?
    b. What are the patterns of interaction as assessed by students' WebCT responsiveness (i.e., frequency of reference that a student makes to peers or the content of their postings; frequency of references made to the readings)?
3. What is the quality of students' learning in this course (as gleaned from their perceptions as well as their performance in the final learning activity)?
4. What is the relationship between how service-learning students perceive WebCT interaction (from Q1), how they engage in WebCT (from Q2), and their learning in the course (from Q3)?

IV. Methods

Data Collection: There are four sources for data:

- Field notes: my initial observations of students' reactions to WebCT implementation and students' frequency of participation in face-to-face interaction.
- WebCT log: provides descriptive data on the level of participation (frequency and length) and pattern of interaction (level of responsiveness).
- Interview: provides data on students' perception about WebCT social and intellectual interactions and how it worked for them, as well as their perceived learning in this course.
- Students' performance on the final learning activity: a means to verify students' learning in this course.

Data Analysis: The data will be analyzed in the following manner

- Field notes: These will be used for the following purposes:
    - A source for substantive categories; e.g., "time is an issue," "too much to do," and "silence in class vs. active in WebCT."
    - A source for some interview questions.
    - A source for themes to pay attention to in interviews.
- WebCT log: will be analyzed in the following manner:
    - Level of participation as reflected in a student's frequency and average length of postings.
    - Pattern of interaction, which will be analyzed using an adapted version of Bonk's Online Discussion Coding Guide.
- Students' interview transcripts will be coded according to:
    - Organizational categories: codes anticipated and developed from conceptual framework prior to attending the study site. E.g., "tool for reflection," "collective knowledge," and "time concern."
    - Substantive categories: These help understand what is going on and develop a theory of how this worked for students.
        - Emic: concepts that participants use in making sense of their experience.
        - Theoretical: theoretical terms but inductively inferred from data.
- To verify students' perceived learning in the course, I would check this against their performance in the final learning activity.
- To develop an interpretive understanding of students' perception I would create a conceptual map (using Inspiration) that reflects students' description of their WebCT use and how it worked for them as a learning tool in this course.
- To develop a theory of how WebCT worked for students, I would read students' perception in the light of how they used WebCT:
    - WebCT level and pattern of participation may have implications for how students perceive WebCT interaction.
    - Students' perceptions may explain their level and pattern of participation in WebCT.

V. Validity: "Validity is a goal rather than a product" (Maxwell, 1996, p. 86)

- To deal with "reactivity" I would emphasize to participants that I have no stake in how they perceive this experiment.

- To deal with bias, I should bring to my awareness my personal perception about WebCT and its role in learning, and constantly monitor how this may affect how I analyze data.

Verification Techniques:

- Member-Check; e.g., paraphrase my understanding of something they said and ask them to react to it.
- Reread the transcript to see where it supports or conflicts with my findings (looking for discrepant evidence). Quasi-Statistics?

Generalizability:

- This is a self-selected group. Were the technology-phobic people scared away?
- The impact of the fact that all but one of the participants are women.

EXERCISE 7.1

Developing a Proposal Argument

The purpose of this exercise is for you to outline the *argument* of your proposal, not its detailed content or structure. You want to present the main substantive points that you need to make about your study, and to organize these so there is a clear logic that leads to a justification for the study. These arguments do not have to be developed in the full form that they will have in the proposal itself, but they should provide the essence of the latter, and should form a coherent sequence.

If you are in the beginning stages of planning your proposal, the outline can be very hypothetical and tentative; the purpose of the exercise is for you to start working on developing your argument, not for you to commit yourself to anything. At this point, it's not important whether you have any evidence or citations to back up your claims; after you have developed an outline of your argument, you can then assess where the holes are in your logic and evidence, and what you need to do to fill them in. This is a "come-as-you-are party"; construct the best argument you can with your present knowledge.

You should address all of the issues listed, although not necessarily in the order presented—sometimes explaining your research relationships depends on knowing your methods or setting, and sometimes the reverse. Don't try to write well-developed prose at this stage; bulleted points will be easier to do and more useful for this exercise.

1. *Research goals.* What intellectual, practical, and personal goals will doing this study accomplish, or attempt to accomplish? What problem(s) will the study address, and why is it important to address this (if this isn't obvious)?
2. *Conceptual framework.* What are the most important theories, ideas, and knowledge (personal as well as research) that inform this study? *How* have these shaped the study? What is your own conceptual framework for the study, and how does it use and incorporate these? What do we *not* know that your study will address?
3. *Research questions.* What do you want to learn by doing this study? How (if it isn't obvious) will answering these questions address the study's goals? How are the questions connected to your conceptual framework?
4. *Research relationships.* What sorts of research relationships do you plan to establish with the participants in your study or setting, or with those controlling access to your setting or data, and why? How will you go about this, and how will this be influenced by any existing relationships you have with them?
5. *Site and participant selection.* What setting(s) will you study, and/or what individuals will you include in your study? (If you haven't made these decisions yet, explain how you expect to make them, along with the criteria you plan to use.) What theoretical and practical considerations have influenced these choices? How are these choices connected to your research questions (if this isn't obvious)?
6. *Data collection.* How do you plan to collect your data, and what data will you collect? How will these data enable you to answer your research questions (if this isn't obvious)?
7. *Data analysis.* What strategies and techniques will you use to make sense of your data? Why have you chosen these? Indicate the kinds of analyses you plan to do; don't just give boilerplate descriptions of methods.
8. *Validity.* What do you see as the most important potential threats to the validity of your conclusions? What will you do to address these? What limitations on generalizability do you see?

Harry Wolcott provided a useful metaphor to keep in mind as you develop your proposal:

> Some of the best advice I've ever seen for writers happened to be included with the directions I found for assembling a new wheelbarrow: *Make sure all parts are properly in place before tightening.* (1990, p. 47)

Like a wheelbarrow, your proposal not only needs to have all the required parts, it also has to *work*—to be put together so that it functions smoothly and conveys to others your research design and the justification for this. This requires attention to the connections between the different parts of the proposal, and to how well the proposal, as a written document, can be understood by your intended audience. As described previously, these are two aspects of what I call "coherence." A coherent proposal depends on a coherent design, but it also needs its own coherence, to flow clearly from beginning to end without gaps, obscurities, confusing transitions, or red herrings. As I've emphasized, there isn't one right way to do this; I've tried to give you the tools that will enable you to put together *a way* that works for you and your research.

NOTES

1. This doesn't mean that you should *conceal* your political views; these are an appropriate part of the discussion of your goals, and may be a possible validity threat that you want to address. However, the discussion should focus on how these views inform your design, rather than being a political polemic or irrelevant self-display.
2. Locke et al. (2000, pp. 63–78) provided an excellent discussion of the purposes and construction of a literature review. However, the literature review in their example qualitative proposal is longer than necessary for most audiences, and less coherent than you want yours to be.
3. The term "methodology" is often used for this section of a proposal. Despite its prevalence, this is an inaccurate and pretentious usage, a good example of what Becker called "classy writing." Methodology is the theory or analysis of methods, not what you actually do in a particular study. The *Publication Manual of the American Psychological Association* (2001, pp. 17–20), a commonly used guide for both dissertations and research publications, uses the term "method" for this section of a manuscript, not "methodology."
4. For some suggestions on how to justify a qualitative study to a potentially ignorant or hostile audience, see Maxwell (1993).
5. Locke et al. (2000, pp. 149–199) and Robson (2002, pp. 526–533) discussed the specific requirements of funding proposals.

Appendix

An Example of a Qualitative Proposal

No single proposal can adequately represent the diversity of qualitative research designs and ways of communicating these. If space permitted, I would include two or three proposals here, to emphasize that there is no one right way to structure either a qualitative study or a proposal. Because I can present only one, I've chosen Martha Regan-Smith's proposal for her dissertation, a study of exemplary medical school teachers. Not only does it give a clear, straightforward explanation and justification for the proposed study, but it also raises many of the key issues that most qualitative proposals will have to address. In my commentary, I try to identify and clarify the connections between these issues and my model of research design, and to present alternative ways of handling these issues. The proposal appears here just as Martha submitted it, with only a few additions (marked by brackets) or corrections of typos or punctuation for greater clarity; the appendixes have been omitted, but are listed in the table of contents.

The most serious danger in presenting an exemplary proposal such as this is that you might use it as a *template* for your own proposal, borrowing its structure and language and simply "filling in the blanks" with your own study. This is a sure recipe for disaster. Your proposal needs to fit the study that you are proposing, and an argument that works well for one study may totally fail to justify a different study. Construct your proposal around your *own* design, not someone else's.

HOW BASIC SCIENCE TEACHERS HELP MEDICAL STUDENTS LEARN

The Students' Perspective

Dissertation Proposal
Martha G. Regan-Smith
March 6, 1991
Harvard Graduate School of Education

Abstract

Medical school consists of two years of basic science and two years of clinical training. The sciences taught in the first two years include Anatomy, Biochemistry, Physiology, Pathology, Microbiology, and Pharmacology. As a result of the biomedical information expansion which has occurred in the last eighty years with no increase in the time available to teach this information, the teaching of basic science has become content heavy. In addition, the teaching has become increasingly rapid paced as most schools over the past twenty years have decreased the number of hours spent in laboratories and demonstrations while increasing reliance on lecturing as the way to teach. Medical student performance on the basic science examinations used for licensure has decreased, and, as a result, medical school faculty feel medical student learning of basic science is less than desired.

As a member of medical school faculties for eighteen years, I want to improve medical student learning of basic science by improving the teaching of basic science in medical school. No qualitative studies of basic science teaching in medical school exist. What works for student learning and how it works is not known. In order to understand how teachers can help medical students learn basic science, I propose to do a qualitative study of four exceptional basic science teachers to answer the following research questions: How do these basic science teachers help medical students learn? What do these teachers do to help students learn? How and why do these techniques help students learn? What motivates the teachers to do what they do? Is what students feel teachers do to help them learn what teachers intend? How do student understandings of what helps them learn differ from teacher understandings?

Each of the four teachers studied teaches a different basic science at a typical private medical school in the northeastern United States. The school has a traditional curriculum in which the two years of basic science is taught predominantly using the lecture format. Each teacher is a winner of the student-selected "Best Teacher Award" and each teacher uses the lecture format for his teaching.

Participant observation of the teacher's lectures and teacher and student interviews are the primary data sources. Classes, in addition, are audiotaped for transcription and videotaped. Videotapes are analyzed as well as used as prompts for dialogue when shown to teacher or students. Interviews are tape-recorded, transcribed and coded. Analytic memos are written and coded for each class observation and interview. Matrices are constructed to identify themes and to check evolving concepts. Both teacher and student collaboration is obtained by getting their opinions of my analysis and conclusions. Each teacher's teaching is analyzed separately followed by comparative analysis of all four teachers' teaching. Generated theory will be compared to existing theory which is primarily based in other educational settings or on personal experience. The goal is to identify teaching techniques and behaviors that help students learn and to gain understanding of how and why these techniques help students learn. This knowledge about practice in context can be taught to teachers in faculty development workshops designed to teach teachers how to improve their teaching. By teaching teachers how to better help students learn, it is hoped improved student learning will result.

This abstract is a concise summary, not just of the components of the research design, but of the connections between these—the *argument* of the proposal. Standards and requirements for abstracts vary, and this one is relatively long. However, conveying the basic argument of your proposal should be a major goal of your abstract, regardless of the length.

Table of Contents

Validity Issues
Ethical Issues
Preliminary Findings
Appendixes

Introduction

Since the Flexner Report in 1910, the four year medical school curriculum has comprised two years of teaching the sciences basic to medicine followed by two years of training in the clinical disciplines. The basic sciences include Anatomy, Microbiology, Biochemistry, Pharmacology, Pathology, and Physiology, and the clinical disciplines include Surgery, Medicine, Pediatrics, Psychiatry and Obstetrics/Gynecology. Because of the information explosion in biomedical science during the past eighty years, the basic science curriculum has become "overstuffed" (Eichna, 1980). Usually three to four sciences are taught simultaneously, using predominantly the lecture format. As a result, students are in class 25–33 hours per week throughout the first two years of medical school. This, combined with the student perception of ineffective teaching (Eichna, 1980; Jonas, 1978; Konner, 1987; Awbrey, 1985), has led to student disillusionment with science (Eichna, 1980) and student cynicism about the educational process (Petersdorf, 1987). In addition, the national failure rate on the basic science portion of the National Board of Medical Examiners examinations has risen over the past six years (NBME letter to Deans, Appendix A) without a demonstrable decrease in student undergraduate grade point averages or admission examination scores.

In an effort to improve the teaching of basic science in medical school, I want to study what teachers of basic science actually do to help medical students learn. I propose to conduct a qualitative study of four exceptional basic science teachers' teaching, from the students' perspective, to answer the question, "How do these teachers help medical students learn?" The goal is to identify teaching techniques and behaviors which help students learn, which can then be taught to teachers in faculty development workshops designed to teach teachers how to improve their teaching and hence better assist student learning.

In this brief introduction, Martha sets the stage for what follows by concisely presenting the practical problem that motivates the study and the historical context of this problem (first paragraph), and briefly stating the goals and nature of the proposed study (second paragraph). The abstract has already given some information about the problem and the study, and further details are left for later. Different studies will require different amounts of information in order to adequately orient the reader to the research problem and study.

Conceptual Framework

To increase medical student enthusiasm for and learning of basic science, several scholars have called for critical examination of the teaching of basic sciences (Bishop, 1984; Neame, 1984; Beaty, 1990). A small number of schools, such as McMaster and Harvard, have been able to replace lectures with small group tutorials during which students participate in problem-based learning by independently solving paper patient cases (Neufeld and Barrows, 1974; Schmidt, 1983). Most medical schools, however, because of financial and faculty constraints, must continue to rely on lectures as a major method of teaching basic sciences. Therefore investigation of how the lecture method can be effective in assisting student learning is worthwhile.

> This paragraph justifies studying the lecture method of teaching basic science. It works well here, but it could also have been included in the introduction.

Existing Literature on Basic Science Teaching in Medical School

Studies of science teaching in secondary or undergraduate schools do not necessarily apply to the medical school setting. The teaching of science through the use of lectures in medical school is unlike the teaching of science in any other educational setting. The rapid pace of medical school and the vast quantity of material needed to be learned by students with varying science backgrounds makes the teaching of science and the learning by the students unique. Effective teaching through the use of lectures in nonmedical school educational settings has been well described and studied (Katona, 1940; McKeachie, 1969; Hyman, 1974; Eble, 1976), but whether the teaching techniques recommended are appropriate in the medical school setting or whether other techniques are helpful is unknown. Qualitative study asking students what works for their learning is needed.

The medical education and health professions education literature on lecturing is limited. Some prescriptive works on how to give effective lectures (Miller, 1962; Bughman, 1973) are based on implicit theory derived from personal experience as students and as faculty (Cook, 1989). Others have been written by educators working in the medical school arena (Jason, 1982), but these are based on educational theory derived from educational settings other than medical school. Schwenk and Whitman (1987) prescribe effective lecturing techniques related to existing educational theory and relate these techniques to communication theory and negotiation theory inherent in effective doctor/patient relationships.

Quantitative studies of lecturing in medical school, usually utilizing student ratings of lecturing techniques, depend on the researchers' prior understanding and assumptions about what helps students learn. Because no qualitative studies of medical student learning of basic science exist, this understanding is based on theory derived from study of or experience with nonmedical school settings. The few quantitative studies in the literature looking at basic science teaching in medical school (Naftulin, 1973; Ware, 1975; Mendez, 1984; Russell, 1984) are limited in scope and contribute little to the research question, "How do basic science teachers help medical students learn?"

Naftulin (1973), looking at teaching delivered in a "seductive charismatic manner," showed that students could give high ratings of such teaching, however, the audience's perception of learning was not included in the study. In response, Ware (1975) concluded that "seductive, charismatic lecturers" assist student learning by showing that students attending lectures with high seduction (characterized by enthusiasm, humor, friendliness, expressiveness, charisma, personality) and low content have similar examination scores as students attending low seduction high content lectures. How these teacher characteristics contribute to student learning of content was not addressed. Mendez (1984) surveyed year I and II medical students for the factors contributing to lecture attendance and found that students attend lectures which they perceive to have clearly defined objectives and which covered material tested on the final examination. How the objectives help student learning and which lecture techniques helped learning were not investigated. Russell (1984) looked at medical student retention of basic material immediately after and fifteen days following lectures with varying amounts of content and found that increasing information density of lectures reduced retention of the basic information. The reasons for this effect were not a part of the study.

Slotnick (1975) and Irby (1976), using quantitative methods, demonstrated that teaching criteria presumed by the researchers to be important for student learning were in fact important to students for their learning. Slotnick (1975) showed that faculty-student rapport, student work required outside of class, pace of class, overall workload, understandability of lecture material, lecturing activities (e.g. summarizes material, concise explanation, organization of material in a logical way), student ability to organize material, and professor knowledge of students' knowledge level are interrelated rather than univariate factors in effective teaching. How these factors affect student learning and why was not a part of the study. Irby (1976) showed that teachers could improve their teaching when given immediate feedback about student ratings of their teaching. The rating variables were derived from education literature and whether the list of teaching techniques rated by the students included all the techniques helpful for student learning could not be determined from the study.

No one has asked medical students what teachers do to help them learn. Existing research has asked students to rate particular teaching techniques or to state whether a technique works or not. These studies depend on the researchers' understanding of what works for student learning. What works to help students learn science in other educational settings may not work in medical school. Quite possibly basic science teachers in medical school have happened upon or developed teaching techniques that are unique to medical school or are unintentionally assisting learning in ways they do not appreciate. Qualitative study is needed to generate a theory of effective nonclinical teaching in medical school.

This section of the proposal argues that we know very little about how basic science teachers in medical school help their students to learn. This point is important in justifying a qualitative study of this phenomenon. As a result, however, the proposal says little about what will be the focus of the conceptual framework section of most proposals: existing theory about, and research on, the phenomenon studied. Martha briefly reviews several theories about what constitutes effective teaching in medical school lectures, but her main point is that these studies address neither *how* such teaching methods work nor the students' perspective. If your study is of a topic for which there exists a substantial literature of theory and research, your conceptual framework section will need to address this literature, as well as your own experience (which Martha discusses in the next section) and pilot research (which she deals with both in the next section and later, in the preliminary findings section).

Personal Interest

I am a physician, an internist and rheumatologist. I was a chemistry major in college, and, prior to this study, I had not participated in a science class since I was a medical student 21 years ago. I have taught how to diagnose adult disease in clinical medicine for 18 years. Approximately six years ago I realized I was also trying to teach both critical thinking skills and the communication skills needed to enable others to understand the reasoning behind a diagnosis. I also realized that I did not know much about critical thinking or communication, let alone how to effectively teach these skills. In 1987 I entered the [Harvard Graduate School of Education] master's program to learn about these skills and how they can be taught. I felt these skills should be a part of a physician's education, and I quickly learned that effective learning of these skills necessitated teaching of these skills throughout medical school, not just in clinical medicine courses.

In 1988, for a course on perspectives of teaching, I was required to study a teacher, classroom or school. I chose to study a teacher. As the Assistant Dean for Clinical Education, whose responsibility is to oversee all clinical teaching, I anticipated I could more easily gain entry into a teacher's classroom if I chose to study a basic science teacher rather than a clinical teacher. In addition, I chose to study a winner of the student awarded "Best Teacher Award." I reasoned that I could learn more about teaching from a winner of such an accolade than a nonwinner, and that a winner would be more likely (i.e. have more confidence) to allow my presence in his classroom than a nonwinner.

I expected the teacher to be skillful; however, I was awed by the extent of his skill as a teacher. Equally surprising was how articulate the students were at describing how he helped them learn. Although I appreciated how he helped me learn in the classroom, I needed student input to appreciate all the aspects of what he did and why it worked for them. Curiosity about how other teachers help medical students learn basic science, and my desire to improve medical education, led to my application in 1988 to the doctoral program with plans to pursue study of how basic science teachers help students learn. By finding out, from the students' perspective, what works to help students learn, I want to discover how teachers can help their students learn and why. Two more teachers have been studied as part of methods courses: the most recent was written up as my qualifying paper entitled "Relevance in Teaching." Each teacher has exemplified all the teaching characteristics that I identified as helping students learn; however, each teacher has best exemplified a different teaching characteristic. The information gleaned from these teacher studies can be used in faculty development workshops designed to teach teachers how to better help their students learn.

> In this section, Martha describes how the study originated, presenting her personal goals and how these connect to the practical and theoretical goals described in the introduction. She also describes her own background as the "research instrument" of the study. In doing these, she also begins to build her justification for the selection of exemplary teachers as the focus of the study, and for using students as a major source of data.

Proposed Research

Research Goals

I want to learn what teachers do to help students learn. The teaching techniques gleaned from teachers in practice which I identify as helping students learn will be useful for other teachers to improve their teaching. Quantitative researchers define the problems of practice in their own terms, not the terms of the practitioners, and tend to generate knowledge that is not useful to the practitioner (Bolster, 1983). Quantitative research often does not cause change in practice, whereas qualitative research, which strives to understand the meaning of action to the participants, can offer improvement of arguments for practice and hence can have greater effect on practice (Fenstermacher, 1986). Knowledge generated by quantitative educational research is often not useful to practitioners who are swayed more by practical arguments, experience and faith (Buchmann, 1984). To improve practice, educational research needs to emphasize the context within which the activities studied occur and the meanings of activities studied for the participants. Qualitative research methods meet these needs (Abrahamson, 1984).

The unique teaching/learning situation in the first two years of medical school merits a qualitative research design which (1) takes into account the contextual elements which makes medical education different from other science education settings and (2) allows for inductive hypothesis generation. What works for basic science lectures is unknown. What helps medical students learn may well be different than what works for students of science in other settings. There is a need for students to define and explain what works. Understanding how particular methods work will require understanding of the context. Using qualitative research methods to study teachers and their students in basic science lecture-format classrooms, I intend to learn from the students and their teachers how basic science teachers help students learn.

For my dissertation I propose to study four basic science teachers. Recognizing that students can be valid, reliable, and useful evaluators of teaching (Costin, 1971; Rippey, 1975; Palchik, 1988; Irby, 1977), I decided to continue to study student selected "Best Teacher Award" winners. I will analyze each teacher's teaching individually, and then comparatively analyze the data collected from all four teacher studies. The theory generated about basic science teaching will be compared to existing effective teaching theory generated from other educational settings.

In this section, Martha reviews the main question and goals of the study, and uses these to justify a qualitative study. In the process, she brings in two additional pieces of the conceptual framework, which relate particularly to methods: the relatively greater impact of qualitative research on practice, and the validity of student ratings of teaching. This discussion could just as easily have been included in the conceptual framework section.

Research Questions

The research questions to be answered are: How do these basic science teachers help their students learn? What do these teachers do to help students learn? How and why do these techniques help students learn? What motivates teachers to do what they do? Is what students feel teachers do to help them learn what teachers intend? How do student understandings of what helps them learn differ from teacher understandings?

In this section, Martha expands on the single main question she stated in the introduction, specifying the range of questions and subquestions that she will address. In many proposals, more explanation or justification of the questions would be desirable, but because of the clear rationale that Martha provides for these questions in previous sections, it seems unnecessary here. For clarity, it is often better to number your research questions, and to indicate which of these are subquestions of particular main questions.

Research Site

I chose to study teachers at a private Northeastern medical school where I have been on the faculty for ten years (I was a winner of the "Best Teacher Award" for clinical teaching in 1987) and I have been the Assistant Dean for Clinical Education for four years. The school is a typical private medical school of slightly less than average student body size. It has a traditional curriculum with two years of basic science followed by two years of clinical experience.

The students are fifty to sixty-five per cent males and thirty-five to fifty per cent females and come from over 50 different public and private schools throughout the United States. Passage of the National Board of Medical Examiners examinations is not required for promotion or graduation; however, most students take the examinations to obtain licensure to practice. The school's matriculating students' admission grade point averages and admission examination scores are near or slightly above the national mean. During the past five years, the school's students' failure rate on the basic science portion of the National Board of Medical Examiners examinations has been at or near the national failure rate and has risen as the national failure rate has. The only differentiating features of this school from other U.S. medical schools are its rural location and its close, friendly faculty/student rapport.

I have professional relationships of considerable mutual respect with the teachers I have chosen to study. All have worked with me as colleagues on Dean's Advisory, Curriculum and/or Student Performance Committees. We see each other as education advocates in an environment which does not reward education program development or teaching achievement. The four teachers chosen from the "Best Teacher" list to be studied each teach at least twenty hours of different basic science discipline courses (Appendix B) and primarily use the lecture format. The basic science teacher winners that will not be studied either teach the same discipline as another studied teacher or teach using a non-lecture method (see Appendix B).

Three teacher observations and interviews have been completed. The teacher remaining to be studied is to be included because he has passion for his subject, which is a recognized dimension of effective teaching (Eble, 1976). Students participating in my previous studies of medical school basic science teaching have recommended study of this professor, who teaches Pathology, because they perceive him as best exemplifying love of subject, which they feel is very important for their learning.

In this section, Martha accomplishes two purposes. First, she describes the setting of her proposed study (supporting the generalizability of her results) and the kind of study she plans to do, and further justifies her choice of teachers. Second, she explains some aspects of her research relationship with the teachers she will be studying. The proposal would have been stronger if she had said more about this, and about her relationship with the students.

Methods of Data Collection

Qualitative research methods were selected for this study both because I did not know a priori what I would find, and because I wanted to generate data rich in detail and embedded in context. Classroom participant observation, student interviews, and teacher interviews are the primary sources [methods] of data collection. In addition, course outlines, syllabi, quizzes, examinations and examination results, paper cases, slides, and other handouts are collected as data. Student evaluations of the course and of the teacher's teaching are also used if available.

For all case studies I attend all possible scheduled lectures given by the teacher throughout a four month course. This will be no less than 2/3's of the teacher's teaching. Two to four lectures are audiotaped to record exactly what was said by the teacher and students in the classroom and later transcribed. As discussed below I videotape teachers teaching and interview both students and teachers. I take field notes while in class unless I am videotaping, and write analytic memos and contact summaries (Miles and Huberman, 1984) following each class as well as each interview.

These two paragraphs provide an overview of the methods section as a whole, and explain the selection strategy for her observations. The selection of students is dealt with later, in the student interviews section.

Videotaping

Videotaping, which I first used with the third teacher I observed, produces a rich source of data about what is going on in the classroom. It allows me to see things I could not see otherwise. I will have the opportunity to review classroom action and observe and isolate individual parts of what is going on. Several of the videotapes will also be used to facilitate the teacher discussing his own teaching in depth. By showing the teacher the tapes of his teaching, I can ask about individual components of his teaching in context. In addition, the tapes will be used to stimulate student dialogue. They will be shown to students to facilitate their explaining the effect of what the teacher does in the classroom to help their learning. Since videotaping was not used to study all four teachers, a comparative analysis cannot be done including all teachers.

Note that videotaping serves two different purposes in this study: ensuring the descriptive validity of her observations, and stimulating recall and reflection as a component of some of the interviews with teachers and students. Videotaping only two of the four teachers would be a serious flaw if the primary purpose of this study were to compare the teachers, but the purpose is to obtain an in-depth understanding of each of the four teachers, and it would be pointless to forego the advantages of videotaping the last two teachers simply to maintain a superficial consistency of method. In a proposal that will be reviewed by readers not familiar with qualitative research, such a decision might need more explicit justification.

Student Interviews

The student interviews begin with an open-ended question such as "What stands out for you?" or "What did you notice?" Subsequent questions are conversational in an attempt to get the interviewee to discuss further something he/she mentioned in an answer. For the first several interviews, the only other preconceived question is "What does the teacher do that helps you learn?" As I observe more classes, questions arise for which I need answers in order to confirm my observation conclusions and to understand what is going on in the classroom, and these are added. Eventually a set of questions (Appendix C) emerges from the evolving data; I ask these questions of all remaining interviewees in addition to the two original set questions.

Out of a class of 84 students, ten to twenty formal student interviews, lasting 20–45 minutes each, are conducted for each teacher study. As many of the student interviews as time will allow are done after the final examination to minimize student fear that what they say will affect their grade. The interviews occur in my office, and are audiotaped and later transcribed. Each interview is preceded by my stating that I am studying what teachers do in the classroom to help students learn, and all interviews are kept anonymous. Analytic memos and contact summary sheets discussing setting, student attitude and demeanor, and content are written for each interview.

The students I interview are selected to contribute student opinion and characteristics that seem important to the context of the study. In the three concluded studies and planned for the fourth study, I seek samples of the student population guided by my emerging theory using theoretical sampling (Strauss, 1987). I do not attempt to get an empirically "representative" sample. As I learn about and make sense of the events in the classroom and its meaning to the participants, I look for negative data as well as positive data for my emerging theory. I determine how many interviews I will do by doing interviews until I find that I am discovering nothing new. I purposely interview students known to be outspoken and critical to be sure I hear negative comments, as well as students known to be outsiders (loners not a member of one of the cliques in the class) to be sure to get different opinions rather than just "the party line." By asking interviewees to tell me who in the class has opinions about class and the teacher different from their own, I find which students are likely to provide contrasting perspectives. In addition, I try to interview students who do not regularly attend class in an effort to understand what informs their decisions to attend or not to attend class.

In this section, Martha presents and justifies both her selection strategy for the student interviews and how she will conduct these. Again, the lack of uniformity of interview questions for all students would be a flaw if the purpose of the study were to compare student responses, but it is not. The number of student interviews could have received more explicit justification, but most readers would feel that this is a more than adequate number. Further justification for her selection decisions is provided in her discussion of validity, and these decisions are supported by her preliminary results.

Teacher Interviews

For all four studies, the teacher is interviewed formally three to six times, and all interviews are audiotaped and transcribed. The interviews occur throughout the course as well as after the course if appropriate. In general, the interview questions are about issues about which I become curious as an observer in class or as the result of student input. I pursue issues raised by the teacher, and ask preconceived questions only if the teacher does not spontaneously address an issue of interest to me.

Formal teacher interviews last at least 30–55 minutes. For two of the teachers, I will use a class videotape as "text for dialogue" about the teacher's teaching for at least one interview. This yields more specific information about the teacher's play-by-play reasoning and strategy than interviews without videotapes, which tend to yield more abstract general teaching strategies and attitudes. Data gathered is analyzed along with the class observations in daily analytic memos and contact sheet summaries.

Because Martha had already collected much of her data when she wrote this proposal, she has a dilemma with what tense to use. Her decision to use mostly present tense seems to be the best choice; this could be misleading, but she has clearly explained earlier that she has already completed data collection for three of the four teachers. For dissertation proposals, I advise you to be completely candid about how much of your data you have already collected, unless you receive authoritative advice to the contrary.

Methods of Analysis

Single-Case Analysis

Analysis of collected data is ongoing. Analysis of transcribed interviews and classes is coded during data collection as soon as transcriptions are available. Codes are inductively generated using the "grounded" approach of Glaser (1965) and emerge from the participants' descriptions of the teacher's teaching. In addition, coding is done using codes from a "start list" (Miles and Huberman, 1984) generated from previous studies. All interviews and classroom transcripts are reread specifically for codes which emerge from later interviews. As patterns or themes are identified, dimensionalization (Strauss and Corbin, 1990) is carried out accompanied by recoding for the developed dimensions or properties of a given theme.

Matrices are constructed from the data and are used to identify patterns, comparisons, trends, and paradoxes. Further questions and possible routes of inquiry are devised to answer the questions which emerge from matrices. Periodic review of all the collected data as well as all the analytic memos followed by summary construction and formulation of yet to be answered questions is done every two or three weeks throughout the study. In addition, I meet weekly with an education colleague, knowledgeable about qualitative research and the research site, to summarize the status of the research and to discuss emerging themes, concepts and explanations.

In the final phase of data analysis each interview is reread with the objective of writing individual short interview summaries. These summaries allow me to see threads that run through interviews and thereby maintain the context for the quotes which are lifted out of the interviews and used as examples in writing up the research. Using Microsoft Word (Apple, 1988), I then cut and paste quotes from all the interviews creating new separate documents for each code that had emerged from analysis of the interviews. This compilation of quotes for each code is used to appreciate trends, contrasts, and similarities. Matrices are constructed to check the validity of themes which emerge. Finally the data is reviewed to pair up student perspectives with teacher perspectives of the same phenomenon to compare and contrast perspectives as well as to look at whether what the teacher intends is, in fact, what the students perceive as happening.

Validation of data is achieved by triangulation (Denzin, 1970) of methods by comparing student perspectives, teacher perspective, and participant observer perspective of events in the classroom. Theoretical validation is achieved by regular presentation and discussion of emerging conclusions with medical school colleagues familiar with the setting, students and teachers. Further validation is achieved by discussing my analyses and conclusions with the teacher and with students.

Cross-Case Analysis

Once I develop an understanding about how the fourth teacher helps his students learn, I will begin cross-case analysis. The first step will be construction of a conceptual framework (Miles and Huberman, 1984) containing the dominant themes of how these four teachers help students learn. Each theme will be dimensionalized (Strauss and Corbin, 1990) or broken into factors and graphically displayed illustrating the relationships between them.

Patterns and themes will be sought by construction of cross-case displays and matrices. Plausible explanations and metaphors will emerge as the variables are related, split and factored (Miles and Huberman, 1984). The goal will be to build a logical chain of evidence (Scriven, 1974) and to construct a theoretically and conceptually coherent theory by checking for rival explanations and looking for negative evidence. In order to check for theory validation informants will be asked for feedback on generated theory after data collection is completed.

Martha's description of her analysis strategies is detailed and comprehensive, but rather abstract and "boilerplate," and doesn't give a good sense of the actual methods and coding categories she'll use; examples would be helpful here. This weakness is remedied by her discussion of preliminary findings, below, which provides detailed, concrete examples of the *content* of her analysis. Her discussion of evidence, rival explanations, and feedback also paves the way for the next section, on validity. In this section, she also tends to slide into impersonal, passive-voice language, which seems incongruent with the mostly first-person, active-voice language of previous sections.

Validity Issues

1. Teacher selection: After the fourth teacher study, I will have studied the award winners from four different discipline courses who use the lecture method (Appendix B). I will stop at four teachers, unless another important teaching characteristic is identified that I have not already found. Because the study school has no features which make it different from other U.S. medical schools with a traditional curriculum of two years of basic science and two years of clinical experience, I find no reason to study teachers elsewhere. Most teachers of basic science in most schools are male, so I found no validity threat to my study by the teachers being male.

This is really an argument for the *generalizability* of her results, not their validity.

2. Student selection: Did I interview enough students? Did I bias the data by who I interviewed? I intentionally try to interview students who have different perspectives and opinions of the teacher's teaching. I interview students who are: (l) known to be outspokenly critical of teaching, (2) from all quartiles of the class, (3) from a variety of career choices, (4) whom I know and whom I barely know, (5) who are referred to me by classmates as feeling differently about the class and teacher, (6) who participate in the typical camaraderie of the class and those who do not, and (7) who attend most every class and those who attend only a few. In essence, I try to seek out students who do not feel the teacher helps them learn as well as those who do. Thereby I try to get both negative and positive student input. I stop interviewing when I begin to hear the same things repeated and no new information.

This paragraph deals with some plausible threats to the validity of her results. The selection strategy described here is an example of purposeful selection; the decision on when to stop interviewing is based on what Strauss (1987) called "theoretical saturation."

3. How do I know what students say is true and not just what I want to hear (i.e. that the teacher helped them learn when he did not)? To make students comfortable being honest with me I assure the students anonymity and interview them in a location distant from the classroom. As often as possible I postpone student interviews until after student grades have been awarded. I also attempt to interview students who are scheduled to finish their third and fourth years at another medical school, thereby eliminating any power I may have as Dean for Clinical Education over them. In the three completed studies, students have not held back from criticizing the teachers nor sharing with me their negative feelings and opinions of the teachers' teaching. I use my presence in the classroom as a learner trying to understand new subjects (e.g. the molecular biology of viruses) to substantiate whether a teacher truly helps students learn. If the teacher helps me learn and the students said he helps them and they pass the course, I believe them. I ask students to give examples of all teaching characteristics they claim help them learn and then I substantiate student examples by being present in class. Collaboration with students (both those in the study and those who were not) by discussing my observations and my conclusions also helps increase my confidence in the validity of my work.

This paragraph addresses her relationship with the students, which has ethical as well as validity implications, and argues that her relationship to them as Dean is not a validity threat to her conclusions. Someone who didn't know Martha and her reputation among these students might not find this argument completely convincing, but I'm not sure what else she could say. The most persuasive point, for me, is that the students she's interviewed *have* been critical of their teachers.

4. How do I know what the teacher says he does is true? I substantiate all teacher claims by participant observation and through student interviews. Teacher beliefs and stated reasons for behavior are accepted as true unless I encounter discrepant evidence.

Here, Martha basically relies on triangulation to deal with the validity threat of self-report bias in the teacher interviews. She could also have used the argument she made in discussing the student interviews: that, having already studied three of the teachers, she *knows* that the observations and student interviews corroborate the teachers' reports.

This section as a whole is organized by particular validity threats—how she might be wrong. In discussing these threats, Martha draws on information previously presented in the methods section, but she reorganizes this information so that it is clear how the data obtained through these methods will help her to deal with these threats.

Ethical Issues

Could my research harm the students or teachers? The teachers risk my finding out that they are not as good a teacher as their award recipience would merit. Even though I do not oversee the basic science part of the curriculum, my administrative colleagues do; and I am a member of the Curriculum Committee. To minimize this fear of risk, each teacher is assured that no one other than specified study school education colleagues with whom I discuss results and conclusions (and my thesis readers) will know of the results of my research unless the teacher gives me permission to do otherwise. I can not eliminate this risk for the teachers.

No harm from teachers can come to the students who participate because the students' identities are kept secret. I can not eliminate the risk that I as the Dean, who writes the student's letter of recommendation for residency after graduation, will form opinions about them as a result of my interview. Those students concerned about such a risk can easily avoid participation. I am aware of no one refusing to participate when asked, hence I do not think student avoidance of participation poses a significant validity threat to my research.

This section could be placed either before or after the validity section. One point that could have been made explicitly here is that these teachers, as award winners, have less to fear from examination of their teaching than most teachers. Martha could also have dealt more convincingly with the ethical issue of risk to the students. Ultimately, her argument depends on her own integrity. The point at the end, about validity, belongs in the previous section.

Preliminary Findings

To date, preliminary analysis of the data has enabled me to identify a number of teaching characteristics which help students learn: clarity, relevance, knowledge of students' understanding, teaching to different learning styles, and passion for the subject. Each of the three teachers studied so far has been found to best exemplify different teaching characteristics even though the characteristic was found in all the other teachers' teaching. In other words, the characteristics identified that help medical students learn basic science are practiced by all the teachers studied but each teacher is a "master" at one or two different characteristics.

The first teacher teaches heart physiology, anatomy and clinical disease to Year II students as a part of the Scientific Basis of Medicine course. The students felt that his lecture style was "like a conversation" with them; the students felt he understood what they knew and what they did not. In addition, this teacher addressed multiple student learning styles by presenting the course material (e.g. coronary artery disease) in seven different ways (i.e. lecture, reading assignments with clear stated objectives, computer interactive patient cases, student participation in demonstrations, small group discussions, problem solving of paper cases, and student presentations of current articles to small groups).

The second teacher teaches the virology section of the Microbiology course in Year I. The students and the teacher felt that the most important feature of his teaching was clarity. The students perceived him to achieve clarity by (1) limiting the material needed to learn, (2) explicitly defining the material the students need and do not need to know, (3) specifying the meaning of his words, (4) presenting concepts moving from the simple to the complex in a logical progression, (5) including stories about patients, epidemiological problems or medical history to explain concepts, (6) asking the class questions critical to understanding the concepts, and (7) repetition of key concepts and facts. He checks on his clarity by giving weekly quizzes and spending extra time in class to explain any quiz questions missed by a significant number of students. The quizzes promote clarity for the students because they additionally give the students feedback on what they know and do not know as well as force them to learn the material weekly and keep up with learning the material rather than cramming for the final examination.

The third teacher teaches pharmacology and best exemplifies the use of relevance in teaching. He uses relevance in his classroom teaching by structuring each lecture around either a presentation of a patient case of his own or a patient case volunteered by a student. In addition, each week he provides students with paper case problems to solve individually thereby letting students simulate practice as physicians. Relevance is also achieved by having students teach students how to solve the case problems. The ensuing class discussion allows students (and the teacher) to learn and discuss student understanding of the pharmacologic principles. The use of the Socratic method by this teacher as cases are discussed in class gives the students opportunity to privately reflect on their own similar experiences with patients. Relevance is also achieved by students privately conversing during class, relating to a neighbor what they are learning in class to cases they have seen, and sharing the experience with the classmate.

Previously studied teachers were not aware of all they did in the classroom to help students learn. Often a teacher is unable to fully appreciate how he helps students learn without my feedback. From the fourth teacher I expect to learn how a teacher's passion for or love of subject helps students learn. I have heard the fourth teacher speak and he is mesmerizing. His charismatic style of presentation captures the audience's attention and, I suppose, it helps them remember what he says. He may also contribute to their learning by motivating them to learn on their own.

I expect the comparative analysis to reveal that the dimensions of each of the individual teacher's teaching characteristics overlap (e.g., anecdotes used to achieve clarity also achieve relevance). Ongoing analysis of my first three case studies reveals that students feel that student-involved teaching, such as students teaching students, is particularly useful for their learning because it achieves clarity, relevance, a form of student/teacher conversation, and addresses student learning styles.

This discussion of preliminary findings serves several purposes. First, it supports Martha's argument that the methods she proposes are workable and will allow her to generate interesting and valid answers to her questions. Second, it fleshes out her rather abstract and general discussion of data analysis, clarifying how she is coding her data and integrating themes within each case, and suggesting issues that the cross-case analysis will focus on.

In summary, by using qualitative research methods to study basic science teachers who primarily use the lecture format to teach, I intend to find how these teachers help medical students learn. The theory generated will be compared to existing theory on effective teaching using lectures in other educational settings. This theory will be used to develop faculty workshops to teach teachers how to teach. The ultimate goal of improved basic science teaching in medical school is to improve medical student enthusiasm for, and learning of, the sciences basic to medicine.

This final paragraph sums up the study by briefly reviewing, in the reverse order from their presentation in the proposal, four components of the design: the methods, the research question, the theoretical framework, and the goals of the study. In doing this, it clearly shows the connections between these components, and links the proposed research to the goals with which the proposal began. However, this is pretty terse for a conclusions section; most proposals will need to say more to summarize the proposal and present the implications of the study.

References

Agar, M. (1991). The right brain strikes back. In N. G. Fielding & R. M. Lee (Eds.), *Using computers in qualitative research* (pp. 181–194). Newbury Park, CA: Sage.

American Psychological Association. (2001). *Publication Manual of the American Psychological Association* (5th ed.). Washington, DC: Author.

Atkinson, P. (1992). The ethnography of a medical setting: Reading, writing, and rhetoric. *Qualitative Health Research, 2,* 451–474.

Becker, H. S. (1970). *Sociological work: Method and substance.* Chicago: Aldine.

Becker, H. S. (1986). *Writing for social scientists: How to start and finish your thesis, book, or article.* Chicago: University of Chicago Press.

Becker, H. S. (1991). Generalizing from case studies. In E. Eisner & A. Peshkin (Eds.), *Qualitative inquiry in education: The continuing debate* (pp. 233–242). New York: Teachers College Press.

Becker, H. S., & Geer, B. (1957). Participant observation and interviewing: A comparison. *Human Organization, 16,* 28–32.

Becker, H. S., Geer, B., Hughes, E. C., & Strauss, A. L. (1961). *Boys in white: Student culture in medical school.* Chicago: University of Chicago Press.

Berg, D. N., & Smith, K. K. (1988). *The self in social inquiry: Researching methods.* Newbury Park, CA: Sage.

Bhattacharjea, S. (1994). *Reconciling "public" and "private": Women in the educational bureaucracy in "Sinjabistan" Province, Pakistan.* Unpublished doctoral dissertation, Harvard Graduate School of Education.

Bloor, M. J. (1983). Notes on member validation. In R. M. Emerson (Ed.), *Contemporary field research: A collection of readings* (pp. 156–172). Prospect Heights, IL: Waveland Press.

Bogdan, R. C., & Biklen, S. K. (2003). *Qualitative research for education: An introduction to theory and methods* (4th ed.). Boston: Allyn & Bacon.

Bolster, A. S. (1983). Toward a more effective model of research on teaching. *Harvard Educational Review, 53,* 294–308.

Bosk, C. (1979). *Forgive and remember: Managing medical failure.* Chicago: University of Chicago Press.

Bredo, E., & Feinberg, W. (1982). *Knowledge and values in social and educational research.* Philadelphia: Temple University Press.

Briggs, C. (1986). *Learning how to ask.* Cambridge, UK: Cambridge University Press.

Briggs, J. (1970). *Never in anger: Portrait of an Eskimo family.* Cambridge, MA: Harvard University Press.

Brinberg, D., & McGrath, J. E. (1985). *Validity and the research process.* Beverly Hills, CA: Sage.

Britan, G. M. (1978). Experimental and contextual models of program evaluation. *Evaluation and Program Planning, 1,* 229–234.

Brown, L. M. (Ed.). (1988). *A guide to reading narratives of conflict and choice for self and moral voice.* Cambridge, MA: Harvard University, Center for the Study of Gender, Education, and Human Development.

Bryman, A. (1988). *Quantity and quality in social research.* London: Unwin Hyman.

Burman, E. (2001). Minding the gap: Positivism, psychology, and the politics of qualitative methods. In D. L. Tolman & M. Brydon-Miller (Eds.), *From subjects to subjectivities: A handbook of interpretive and participatory methods* (pp. 259–275). New York: New York University Press.

Campbell, D. T. (1984). Foreword to *Case study research: Design and methods,* by R. Yin. Beverly Hills, CA: Sage.

Campbell, D. T. (1988). *Methodology and epistemology for social science: Selected papers.* Chicago: University of Chicago Press.

Campbell, D. T., & Stanley, J. (1963). Experimental and quasi-experimental designs for research on teaching. In N. L. Gage (Ed.), *Handbook of research on teaching* (pp. 171–246). Chicago: Rand McNally.

Christians, C. G. (2000). Ethics and politics in qualitative research. In N. K. Denzin & Y. S. Lincoln (Eds.), *Handbook of qualitative research* (2nd ed., pp. 133–155). Thousand Oaks, CA: Sage.

Coffey, A., & Atkinson, P. (1996). *Making sense of qualitative data.* Thousand Oaks, CA: Sage.

Cook, T. D., & Campbell, D. T. (1979). *Quasi-experimentation: Design and analysis issues for field settings.* Boston: Houghton Mifflin.

Creswell, J. W. (1994). *Research design: Quantitative and qualitative approaches.* Thousand Oaks, CA: Sage.

Creswell, J. W. (1998). *Qualitative inquiry and research design: Choosing among five traditions.* Thousand Oaks, CA: Sage.

Creswell, J. W. (2002). *Educational research: Planning, conducting, and evaluating quantitative and qualitative research.* Upper Saddle River, NJ: Merrill Prentice Hall.

Croskery, B. (1995). *Swamp leadership: The wisdom of the craft.* Unpublished doctoral dissertation, Harvard Graduate School of Education.

Denzin, N. K. (1970). *The research act.* Chicago: Aldine.

Denzin, N. K., & Lincoln, Y. S. (2000). Introduction: The discipline and practice of qualitative research. In N. K. Denzin & Y. S. Lincoln (Eds.), *Handbook of qualitative research* (2nd ed., pp. 1–28). Thousand Oaks, CA: Sage.

Dexter, L. A. (1970). *Elite and specialized interviewing.* Evanston, IL: Northwestern University Press.

Dey, I. (1993). *Qualitative data analysis: A user-friendly guide for social scientists.* London: Routledge.

Emerson, R. M., Fretz, R. I., & Shaw, L. L. (1995). *Writing ethnographic fieldnotes.* Chicago: University of Chicago Press.

Erickson, F. (1986). Qualitative methods. In M. C. Wittrock (Ed.), *Handbook of research on teaching.* New York: Macmillan.

Erickson, F. (1992). Ethnographic microanalysis of interaction. In M. D. LeCompte, W. L. Millroy, & J. Preissle (Eds.), *The handbook of qualitative research in education* (pp. 201–225). San Diego: Academic Press.

Festinger, L., Riecker, H. W., & Schachter, S. (1956). *When prophecy fails.* Minneapolis: University of Minnesota Press.

Fielding, N., & Fielding, J. (1986). *Linking data.* Beverly Hills, CA: Sage.

Fine, M., Weis, L., Weseen, S., & Wong, L. (2000). For whom? Qualitative research, representations, and social responsibilities. In N. K. Denzin & Y. S. Lincoln (Eds.), *Handbook of qualitative research* (2nd ed., pp. 107–131). Thousand Oaks, CA: Sage.

Freidson, E. (1975). *Doctoring together: A study of professional social control.* Chicago: University of Chicago Press.

Gee, J. P., Michaels, S., & O'Connor, M. C. (1992). Discourse analysis. In M. D. LeCompte, W. L. Millroy, & J. Preissle (Eds.), *The handbook of qualitative research in education* (pp. 227–291). San Diego: Academic Press.

Geertz, C. (1973). *The interpretation of cultures.* New York: Basic Books.

Geertz, C. (1974). "From the native's point of view": On the nature of anthropological understanding. *Bulletin of the American Academy of Arts and Sciences, 28*(1), 27–45.

Glaser, B., & Strauss, A. (1967). *The discovery of grounded theory.* Chicago: Aldine.

Glesne, C. (1999). *Becoming qualitative researchers: An introduction* (2nd ed.). White Plains, NY: Longman.

Glesne, C., & Peshkin, A. (1992). *Becoming qualitative researchers: An introduction.* White Plains, NY: Longman.

Goldenberg, C. (1992). The limits of expectations: A case for case knowledge of teacher expectancy effects. *American Educational Research Journal, 29,* 517–544.

Grady, K. A., & Wallston, B. S. (1988). *Research in health care settings.* Newbury Park, CA: Sage.

Guba, E. G., & Lincoln, Y. S. (1989). *Fourth generation evaluation.* Newbury Park, CA: Sage.

Guilbault, B. (1989). *The families of dependent handicapped adults: A working paper.* Unpublished manuscript.

Hammersley, M. (1992). *What's wrong with ethnography?* London: Routledge.

Hammersley, M., & Atkinson, P. (1995). *Ethnography: Principles in practice* (2nd ed.). London: Routledge.

Hannerz, U. (1992). *Cultural complexity: Studies in the social organization of meaning.* New York: Columbia University Press.

Heider, E. R. (1972). Probability, sampling, and ethnographic method: The case of Dani colour names. *Man, 7,* 448–466.

Heinrich, B. (1979). *Bumblebee economics.* Cambridge, MA: Harvard University Press.

Heinrich, B. (1984). *In a patch of fireweed.* Cambridge, MA: Harvard University Press.

Huberman, A. M. (1989). *La vie des enseignats.* Neuchâtel, Switzerland: Editions Delachaux & Niestlé. (English translation published in the United States in 1993 as *The Lives of Teachers,* New York: Teachers College Press)

Huck, S. W., & Sandler, H. M. (1979). *Rival hypotheses: "Minute mysteries" for the critical thinker.* London: Harper & Row.

Jackson, B. (1987). *Fieldwork.* Urbana: University of Illinois Press.

Janesick, V. J. (1994). The dance of qualitative research design: Metaphor, methodolatry, and meaning. In N. K. Denzin & Y. S. Lincoln (Eds.), *Handbook of qualitative research* (pp. 209–219). Thousand Oaks, CA: Sage.

Jansen, G., & Peshkin, A. (1992). Subjectivity in qualitative research. In M. D. LeCompte, W. L. Millroy, & J. Preissle (Eds.), *The handbook of qualitative research in education* (pp. 681–725). San Diego: Academic Press.

Kaffenberger, C. (1999). *The experience of adolescent cancer survivors and their siblings: The effect on their lives and their relationships.* Unpublished doctoral dissertation, George Mason University.

Kaplan, A. (1964). *The conduct of inquiry.* San Francisco: Chandler.

Kidder, L. H. (1981). Qualitative research and quasi-experimental frameworks. In M. B. Brewer & B. E. Collins (Eds.), *Scientific inquiry and the social sciences.* San Francisco: Jossey-Bass.

Kirk, J., & Miller, M. (1986). *Reliability and validity in qualitative research.* Beverly Hills, CA: Sage.

Kvale, S. (Ed.). (1989). *Issues of validity in qualitative research.* Lund, Sweden: Studentlitteratur.

Kvarning, L.-Å. (October 1993). Raising the *Vasa. Scientific American,* 84–91.

Lather, P. (1993). Fertile obsession: Validity after poststructuralism. *Sociological Quarterly, 34,* 673–693.

Lave, C. A., & March, J. G. (1975). *An introduction to models in the social sciences.* New York: Harper & Row.

Lawrence-Lightfoot, S., & Davis, J. (1997). *The art and science of portraiture.* San Francisco: Jossey-Bass.

LeCompte, M. D., & Preissle, J. (1993). *Ethnography and qualitative design in educational research* (2nd ed.). San Diego: Academic Press.

Light, R. J., & Pillemer, D. B. (1984). *Summing up.* Cambridge, MA: Harvard University Press.

Light, R. J., Singer, J., & Willett, J. (1990). *By design: Conducting research on higher education.* Cambridge, MA: Harvard University Press.

Lincoln, Y. S. (1990). Toward a categorical imperative for qualitative research. In E. Eisner & A. Peshkin (Eds.), *Qualitative inquiry in education: The continuing debate.* New York: Teachers College Press.

Lincoln, Y. S., & Guba, E. G. (1985). *Naturalistic inquiry.* Beverly Hills, CA: Sage.

Linde, C. (1993). *Life stories: The creation of coherence.* New York: Oxford University Press.

L.L.Bean October Classics 1998 catalog. Freeport ME: L.L.Bean.

Locke, L., Silverman, S. J., & Spirduso, W. W. (1998). *Reading and understanding research.* Thousand Oaks, CA: Sage.

Locke, L., Spirduso, W. W., & Silverman, S. J. (1993). *Proposals that work* (3rd ed.). Newbury Park, CA: Sage.

Locke, L., Spirduso, W. W., & Silverman, S. J. (2000). *Proposals that work* (4th ed.). Newbury Park, CA: Sage.

Manning, H. (Ed.). (1960). *Mountaineering: The freedom of the hills.* Seattle: The Mountaineers.

Margolis, J. S. (1990). *Psychology of gender and academic discourse: A comparison between female and male students' experiences talking in a college classroom.* Unpublished doctoral dissertation, Harvard Graduate School of Education.

Marshall, C., & Rossman, G. (1999). *Designing qualitative research* (3rd ed.). Thousand Oaks, CA: Sage.

Maxwell, J. A. (1971). *The development of Plains kinship systems.* Unpublished master's thesis, University of Chicago.

Maxwell, J. A. (1978). The evolution of Plains Indian kin terminologies: A non-reflectionist account. *Plains Anthropologist, 23,* 13–29.

Maxwell, J. A. (1986). *The conceptualization of kinship in an Inuit community.* Unpublished doctoral dissertation, University of Chicago.

Maxwell, J. A. (1992). Understanding and validity in qualitative research. *Harvard Educational Review, 62,* 279–300.

Maxwell, J. A. (1993). Gaining acceptance for qualitative methods from clients, policy-makers, and participants. In D. Fetterman (Ed.), *Speaking the language of power.* London: Falmer Press.

Maxwell, J. A. (1995). Diversity and methodology in a changing world. *Pedagogía, 30,* 32–40.

Maxwell, J. A. (1996). *Qualitative research design: An interactive approach.* Thousand Oaks, CA: Sage.

Maxwell, J. A. (1999). A realist/postmodern concept of culture. In E. L. Cerroni-Long (Ed.), *Anthropological theory in North America* (pp. 143–173). Westport, CT: Bergin & Garvey.

Maxwell, J. A. (2002). Realism and the role of the researcher in qualitative psychology. In M. Kiegelmann (Ed.), *The role of the researcher in qualitative psychology* (pp. 11–30). Tuebingen, Germany: Verlag Ingeborg Huber.

Maxwell, J. A. (2004a). Causal explanation, qualitative research, and scientific inquiry in education. *Educational Researcher, 33*(2), 3–11.

Maxwell, J. A. (2004b). Re-emergent scientism, postmodernism, and dialogue across differences. *Qualitative Inquiry, 10,* 35–41.

Maxwell, J. A. (2004c). Using qualitative methods for causal explanation. *Field Methods, 16*(3), 243–264.

Maxwell, J. A. (n.d.). *Diversity, solidarity, and community.* Unpublished manuscript.

Maxwell, J. A., & Loomis, D. (2002). Mixed methods design: An alternative approach. In A. Tashakkori & C. Teddlie (Eds.), *Handbook of mixed methods in social and behavioral research* (pp. 241–271). Thousand Oaks, CA: Sage.

Maxwell, J. A., & Miller, B. A. (n.d.). *Categorizing and connecting as components of qualitative data analysis.* Unpublished manuscript.

McMillan, J. H., & Schumacher, S. (2001). *Research in education: A conceptual introduction.* New York: Longman.

Menzel, H. (1978). Meaning: Who needs it? In M. Brenner, P. Marsh, & M. Brenner (Eds.), *The social contexts of method* (pp. 140–171). New York: St. Martin's Press.

Merriam, S. (1988). *Case study research in education: A qualitative approach.* San Francisco: Jossey-Bass.

Metzger, M. (June 1993). Playing school or telling the truth? *Harvard Graduate School of Education Alumni Bulletin.*

Miles, M. B., & Huberman, A. M. (1984). *Qualitative data analysis: A sourcebook of new methods.* Beverly Hills, CA: Sage.

Miles, M. B., & Huberman, A. M. (1994). *Qualitative data analysis: An expanded sourcebook* (2nd ed.). Thousand Oaks, CA: Sage.

Mills, C. W. (1959). On intellectual craftsmanship. In C. W. Mills, *The sociological imagination.* London: Oxford University Press.

Mishler, E. G. (1986). *Research interviewing: Context and narrative.* Cambridge, MA: Harvard University Press.

Mohr, L. (1982). *Explaining organizational behavior.* San Francisco: Jossey-Bass.

Norris, S. P. (1983). The inconsistencies at the foundation of construct validation theory. In E. R. House (Ed.), *Philosophy of evaluation* (pp. 53–74). San Francisco: Jossey-Bass.

Novak, J. D., & Gowin, D. B. (1984). *Learning how to learn.* Cambridge, UK: Cambridge University Press.

Patton, M. Q. (1990). *Qualitative evaluation and research methods* (2nd ed.). Newbury Park, CA: Sage.

Patton, M. Q. (2001). *Qualitative research and evaluation methods* (3rd ed.). Thousand Oaks, CA: Sage.

Pawson, R., & Tilley, N. (1997). *Realistic evaluation.* London: Sage.

Pelto, P., & Pelto, G. (1975). Intra-cultural diversity: Some theoretical issues. *American Ethnologist, 2,* 1–18.

Peshkin, A. (1991). *The color of strangers, the color of friends: The play of ethnicity in school and community.* Chicago: University of Chicago Press.

Peters, R. L. (1992). *Getting what you came for: The smart student's guide to earning a master's or a Ph.D.* New York: Noonday Press.

Phillips, D. C. (1987). *Philosophy, science, and social inquiry.* Oxford: Pergamon Press.

Phillips, D. C., & Burbules, N. (2000). *Postpositivism and educational research.* Lanham, MD: Rowman & Littlefield.

Pitman, M. A., & Maxwell, J. A. (1992). Qualitative approaches to evaluation. In M. D. LeCompte, W. L. Millroy, & J. Preissle (Eds.), *The handbook of qualitative research in education* (pp. 729–770). San Diego: Academic Press.

Platt, J. R. (1964). Strong inference. *Science, 146,* 347–353.

Poggie, J. J., Jr. (1972). Toward control in key informant data. *Human Organization, 31,* 23–30.

Przeworski, A., & Salomon, F. (1988). *On the art of writing proposals: Some candid suggestions for applicants to Social Science Research Council competitions.* New York: Social Science Research Council.

Putnam, H. (1987). *The many faces of realism.* LaSalle, IL: Open Court.

Putnam, H. (1990). *Realism with a human face.* Cambridge, MA: Harvard University Press.

Rabinow, P. (1977). *Reflections on fieldwork in Morocco.* Berkeley: University of California Press.

Rabinow, P., & Sullivan, W. M. (1979). *Interpretive social science: A reader.* Berkeley: University of California Press.

Ragin, C. C. (1987). *The comparative method: Moving beyond qualitative and quantitative strategies.* Berkeley: University of California Press.

Reason, P. (1988). Introduction. In P. Reason (Ed.), *Human inquiry in action: Developments in new paradigm research.* Newbury Park, CA: Sage.

Reason, P. (1994). Three approaches to participative inquiry. In N. K. Denzin & Y. S. Lincoln (Eds.), *Handbook of qualitative research* (pp. 324–339). Thousand Oaks, CA: Sage.

Regan-Smith, M. G. (1991). *How basic science teachers help medical students learn.* Unpublished doctoral dissertation, Harvard Graduate School of Education.

Riessman, C. K. (1993). *Narrative analysis.* Newbury Park, CA: Sage.

Robson, C. (2002). *Real world research.* Oxford: Blackwell.

Rudestam, K. E., & Newton, R. R. (1992). *Surviving your dissertation.* Newbury Park, CA: Sage.

Ryle, G. (1949). *The concept of mind.* London: Hutchinson.

Sankoff, G. (1971). Quantitative aspects of sharing and variability in a cognitive model. *Ethnology, 10,* 389–408.

Sayer, A. (1992). *Method in social science: A realist approach* (2nd ed.). London: Routledge.

Schram, T. H. (2003). *Conceptualizing qualitative inquiry.* Upper Saddle River, NJ: Merrill Prentice Hall.

Scriven, M. (1967). The methodology of evaluation. In R. E. Stake (Ed.), *Perspectives of curriculum evaluation.* Chicago: Rand McNally.

Scriven, M. (1991). Beyond formative and summative evaluation. In M. W. McLaughlin & D. C. Phillips (Eds.), *Evaluation and education at quarter century* (pp. 19–64). Chicago: National Society for the Study of Education.

Seidman, I. E. (1998). *Interviewing as qualitative research* (2nd ed.). New York: Teachers College Press.

Shadish, W. R., Cook, T. D., & Campbell, D. T. (2002). *Experimental and quasi-experimental designs for generalized causal inference.* Boston: Houghton Mifflin.

Shavelson, R. J., & Towne, L. (Eds.). (2002). *Scientific research in education.* Washington, DC: National Academy Press.

Shweder, R. A. (Ed.). (1980). *Fallible judgment in behavioral research.* San Francisco: Jossey-Bass.

Smith, M. L., & Shepard, L. A. (1988). Kindergarten readiness and retention: A qualitative study of teachers' beliefs and practices. *American Educational Research Journal, 25,* 307–333.

Spradley, J. (1979). *The ethnographic interview.* New York: Holt, Rinehart & Winston.

Stake, R. (1995). *The art of case study research.* Thousand Oaks, CA: Sage.

Starnes, B. (1990). *"Save one of those high-up jobs for me": Shared decision making in a day care center.* Unpublished doctoral dissertation, Harvard Graduate School of Education.

Strauss, A. (1987). *Qualitative analysis for social scientists.* Cambridge, UK: Cambridge University Press.

Strauss, A. (1995). Notes on the nature and development of general theories. *Qualitative Inquiry, 1,* 7–18.

Strauss, A., & Corbin, J. (1990). *Basics of qualitative research: Grounded theory procedures and techniques.* Newbury Park, CA: Sage.

Strauss, A., & Corbin, J. (1998). *Basics of qualitative research: Techniques and procedures for developing grounded theory* (2nd ed.). Thousand Oaks, CA: Sage.

Tolman, D. L., & Brydon-Miller, M. (2001). *From subjects to subjectivities: A handbook of interpretive and participatory methods.* New York: New York University Press.

Tukey, J. (1962). The future of data analysis. *Annals of Mathematical Statistics, 33,* 1–67.

Webster's Ninth New Collegiate Dictionary. (1984). Springfield, MA: Merriam-Webster.

Weiss, R. S. (1994). *Learning from strangers: The art and method of qualitative interviewing.* New York: Free Press.

Weitzman, E. A., & Miles, M. B. (1995). *Computer programs for qualitative data analysis.* Thousand Oaks, CA: Sage.

Werner, O., & Schoepfle, G. M. (1987). *Systematic fieldwork.* Newbury Park, CA: Sage.

Wievorka, M. (1992). Case studies: History or sociology? In C. C. Ragin & H. S. Becker (Eds.), *What is a case?* (pp. 159–172). Cambridge, UK: Cambridge University Press.

Wolcott, H. F. (1990). *Writing up qualitative research.* Newbury Park, CA: Sage.

Yin, R. K. (1994). *Case study research: Design and methods* (2nd ed.). Thousand Oaks, CA: Sage.

Index

NOTE: Page references followed by *fig* indicate an illustrated figure

About the Author

Joseph A. Maxwell is an Associate Professor in the Graduate School of Education at George Mason University, where he teaches courses on research design and methods and on writing a dissertation proposal. He has published work on qualitative research and evaluation, mixed method research, sociocultural theory, Native American social organization, and medical education. He has also worked extensively in applied settings. He has presented seminars and workshops on teaching qualitative research methods and on using qualitative methods in various applied fields, and has been an invited speaker at conferences and universities in the United States, Puerto Rico, Europe, and China. He has a Ph.D. in anthropology from the University of Chicago.